STRUCTURAL ENGINEERING FOR PROFESSIONAL ENGINEERS' EXAMINATIONS

STRUCTURAL ENGINEERING FOR PROFESSIONAL ENGINEERS' EXAMINATIONS

Including Statics, Mechanics of Materials, and Civil Engineering

MAX KURTZ, P.E.

CONSULTING ENGINEER AND EDUCATOR; MEMBER,
NATIONAL SOCIETY OF PROFESSIONAL ENGINEERS;
AUTHOR, *Comprehensive Structural Design Guide,*
Engineering Economics for Professional Engineers'
Examinations, Framing of Hip and Valley Rafters;
EDITOR-IN-CHIEF, *Handbook of Engineering*
  Economics (in preparation);
CONTRIBUTING AUTHOR, *Standard Handbook of*
  Engineering Calculations

THIRD EDITION

McGRAW-HILL BOOK COMPANY

New York St. Louis San Francisco Auckland Bogotá Düsseldorf
Johannesburg London Madrid Mexico Montreal
New Delhi Panama Paris São Paulo
Singapore Sydney Tokyo Toronto

Library of Congress Cataloging in Publication Data

Kurtz, Max, date.
  Structural engineering for professional
engineers' examinations.

  Bibliography: p.
  Includes index.
  1. Structural engineering—Examinations,
questions, etc. I. Title.
TA638.5.K8   1978   624'.17   76-16787
ISBN 0-07-035657-2

5678910   MUBP   8987654321

The editors for this book were Tyler G. Hicks and Margaret Lamb
and the production supervisor was Teresa F. Leaden.

Printed by The Murray Printing Company and bound by The Book Press.

To my wife, Ruth

THE BUILDERS

All are architects of Fate,
   Working in these walls of Time;
Some with massive deeds and great,
   Some with ornaments of rhyme.

For the structure that we raise,
   Time is with materials filled;
Our to-days and yesterdays
   Are the blocks with which we build.

Build to-day, then, strong and sure,
   With a firm and ample base;
And ascending and secure
   Shall to-morrow find its place.

Thus alone can we attain
   To those turrets, where the eye
Sees the world as one vast plain,
   And one boundless reach of sky.

HENRY WADSWORTH LONGFELLOW

CONTENTS

Preface   xi
Recommendations to the Reader   xv
Abbreviations   xviii

1. Analysis of Force Systems **1**

1-1. Definitions;   1-2. Characteristics of Forces and Force Systems;
1-3. Transformation of Force System;   1-4. Resultant of Force System;
1-5. Conditions of Equilibrium;   1-6. Classification of Forces;   1-7. Structural
Supports and Joints;   1-8. Free-Body Diagrams and Internal Forces;
1-9. Analysis of Statically Determinate Structures;   1-10. Static Friction;
1-11. Belt Friction; Problems

2. Geometrical Properties of an Area **34**

2-1. Objective;   2-2. Statical Moment;   2-3. Resultant of Distributed Force;
2-4. Moment of Inertia;   2-5. Polar Moment of Inertia; Product of Inertia;
2-6. Statical Moment of Partial Area; Problem

3. Analysis of Stress and Strain **48**

3-1. Objective;   3-2. Definitions;   3-3. Stresses on Collinear Planes; Mohr's
Circle;   3-4. Deformation of a Stressed Body;   3-5. Simple Statically

Indeterminate Members; 3-6. Thermal Stresses; 3-7. Thin Circular Hoops; 3-8. Elastic Design and Strength Design; Problems

4. Pressure Vessels—Riveted and Welded Connections **76**

4-1. Stresses in Cylindrical Vessels; 4-2. Deformation of Cylindrical Vessels; 4-3. Prestressed Concrete Pipes; 4-4. Stresses in Riveted Joints; 4-5. Design of Welded Joints; Problems

5. Stresses in Beams. **92**

5-1. Definitions; 5-2. Reactions for Statically Determinate Beams; 5-3. Vertical Shear and Bending Moment; 5-4. Moving-Load Groups; 5-5. Bending Stresses; 5-6. Horizontal Shear and Shearing Stresses; 5-7. Composite Beams; 5-8. Combined Bending and Axial Loading; Problems

6. Design of Steel Beams and Plate Girders **131**

6-1. Introduction; 6-2. Distribution of Vertical Shear and Bending Moment; 6-3. Allowable Bending Stress; 6-4. Cover-Plated Beams; 6-5. Design of Plate Girders; 6-6. Bearing and Intermediate Stiffeners; 6-7. Rivet Pitch in Plate Girders

7. Deflection of Beams—Statically Indeterminate Beams. **148**

7-1. Definitions and Notation; 7-2. Curvature of Elastic Curve; 7-3. Double-Integration Method; 7-4. Calculation of Deflection by Superposition; 7-5. Relative Rigidity of Beams; 7-6. Analysis of Beams with Three Reactions; 7-7. Theorem of Three Moments; 7-8. Beam Design Based on Deflection

8. Torsion **162**

8-1. Stresses in Cylindrical Shaft; 8-2. Deformation of Cylindrical Shaft; 8-3. Statically Indeterminate Shafts; 8-4. Torsion on Rivet Group; 8-5. Beam-Shafts; Problems

9. Steel Members in Tension and Compression **174**

9-1. Design of Tension Members; 9-2. Columns with Axial Loads; 9-3. Investigation of Beam-Column

10. Trusses. **185**

10-1. Stability of the Triangle; 10-2. Definition of a Truss; 10-3. Truss Analysis; 10-4. Graphical Method; 10-5. Considerations in Design of Roof Truss; 10-6. Loading of Bridge Truss; 10-7. Vertical Shear and Bending Moment in Bridge Truss; 10-8. Influence Lines; 10-9. Forces in Bridge Truss under Multiple-Load Systems; Problems

11. Timber Design **207**

11-1. Characteristics of Wood; 11-2. Flexural Members; 11-3. Compression on Oblique Plane; 11-4. Timber Columns

12. Loads on Foundations—Stability of Structures **214**

12-1. Determination of Soil Pressure; 12-2. Eccentrically Loaded Pile Groups; 12-3. Criteria for Stability; 12-4. Gravity Dams; 12-5. Retaining Walls; Problems

13. Properties of Concrete—Reinforced-Concrete Beams **226**

13-1. Introduction; 13-2. Design of Concrete Mixtures; 13-3. Properties of Reinforced-Concrete Beams; 13-4. Notation; 13-5. Requirements in Beam Design; 13-6. Rectangular Beams; 13-7. T Beams; 13-8. Doubly Reinforced Beams; 13-9. Shearing Stress, Bond Stress, and Development Length; 13-10. Reinforcement for Diagonal Tension; 13-11. Continuous Beams

14. Reinforced-Concrete Columns **257**

14-1. Basic Design Concepts; 14-2. Equations for Rectangular Members; 14-3. Interaction Diagrams; 14-4. Eccentrically Loaded Columns

15. Column Footings **267**

16. Fluid Mechanics **273**

16-1. Buoyancy; 16-2. Bernoulli's Theorem; 16-3. Power Associated with Fluid Flow; 16-4. Flow through Orifices; 16-5. Flow of Water over Weirs; 16-6. Flow of Liquid in Pipes; 16-7. Uniform Flow in Open Channels; 16-8. Centrifugal Pumps

17. Surveying and Route Design **289**

17-1. Horizontal Circular Curves; 17-2. Properties of Parabolic Routes; 17-3. Design of Parabolic Routes; 17-4. Sight Distances; 17-5. Volumes of Earthwork; 17-6. Plotting a Closed Traverse; 17-7. Calculation of Areas; 17-8. Differential Leveling; 17-9. Stadia Surveying; 17-10. Field Astronomy

18. Soil Mechanics **322**

18-1. Composition of Soils; 18-2. Transformation of Borrow Material; 18-3. Design of Soil Mixtures; 18-4. Shearing Capacity of Soil; 18-5. Compressibility of Soil

x   *Contents*

19. Water Supply and Sewerage **341**

19-1. Hydraulics of Wells;   19-2. Design of Storm Drains;   19-3. Stabilization of Sewage

Appendix A. Analysis of a Parabolic Arc **351**

Appendix B. Engineering Units in the International System (SI) **355**

Bibliography   364
Index   369

PREFACE

The cardinal objective of this book is to facilitate preparation for the statics, mechanics of materials, and civil engineering sections of the professional engineers' licensing examinations given throughout the United States. The text covers the material required by candidates for both the intern engineer (or engineer-in-training) certificate and the professional engineer's license. In addition to P.E. candidates, the book will prove helpful to applicants for civil service positions, architects preparing for registration examinations, and practicing engineers and architects.

This book helps the reader prepare for the P.E. examinations by presenting a vast array of problems that are characteristic of these examinations. It seeks to simplify the reader's task in every possible way. All terms and symbols are explicitly defined and all relevant equations are conspicuously displayed before the numerical calculations are undertaken. Each step in the solution is carefully explained to ensure that it will be readily understood. Thus, through use of this book the reader can acquire the proficiency in problem-solving that the P.E. examinations require.

However, this book goes far beyond the mere solution of numerical problems, for it offers a concise but thorough review of the engineering principles, concepts, and techniques that underlie each subject. To pass the P.E. examinations, the candidate must have a thorough grasp of engineering principles and the ability to apply them correctly in each situation. The examiners are attempting to appraise the knowledge, judgment, and analytical powers of the candidate, not simply his or her capacity for memorizing a formula or dexterity in using a slide rule or calculator. By devising problems adroitly, they can measure the candidate's engineering skills with great accuracy.

The material in this text has been organized in an orderly manner, with one topic following another in logical sequence. For example, the subject of influence lines is developed in Art. 10-8 and then applied to the analysis of bridge trusses in Art. 10-9. Similarly, the geometrical characteristics of a parabolic route are formulated in Art. 17-2 and then applied to route design in Art. 17-3. The evaluation of shearing stresses in a beam requires the calculation of the statical moment of a partial area, and this subject is therefore explored in Art. 2-6. Thus, as the reader undertakes the solution of a problem he or she already possesses the analytical tools that the solution requires. This systematic arrangement of material makes the reader's study a continuous forward movement rather than a spasmodic one.

A major objective of this text is to give the reader flexibility in solving problems, thus enabling the reader to identify and apply the most direct method of solution where multiple methods exist. This matter is of extreme importance, for the candidate is working against a time limit during the P.E. examinations. The individual who is rigidly confined to a single method of solution is severely handicapped in this respect. To help the reader develop the required flexibility, this text solves many problems by alternative methods. Example 10-1 on the analysis of a truss serves as an illustration.

Surveying problems pertaining to the design of parabolic routes appear copiously in the P.E. examinations. Articles 17-2 and 17-3 explain and demonstrate the variety of ways in which a parabolic route can be plotted, thereby giving the reader extensive practice in solving problems of this type. The problem of designing a route to pass through a given point other than the summit or sag is solved in this text by a method that is far simpler than that presented in other books.

Appendix B offers a detailed study of the International System of units, abbreviated as SI. This material covers the units, symbols, and prefixes that are embodied in this system, and it presents a table of conversion factors designed for quick reference. It also demonstrates the manner in which engineering calculations can readily be converted from the foot-pound-second system to SI.

To enhance the usefulness of this book as a text in a review course or self-study program, 53 problems requiring solution by the reader are presented at the end of various chapters. In all instances, the numerical answer is supplied, and in many instances a hint to the method of solution is also offered.

The third edition of this book applies present design codes in steel, reinforced concrete, and timber, and it is oriented toward current P.E. examinations. The following are some of the topics added in preparing the third edition: transformation of force systems, thin circular hoops, stress analysis by Mohr's circle, theorem of three moments, and statically indeterminate shafts. Several other topics underwent major revision and expansion. To accommodate these additions and expansions, it was necessary to delete the following topics: surveying tapes, moment distribution, design of combined footings, design of cantilever walls, stability of earth slopes, and mechanics of compressible fluids. Readers who wish to include these topics in their study should refer to the books listed in the Bibliography or to the second edition of this book. It was also necessary to delete the material titled "Bending-Moment Diagrams for Moving-Load Groups" and "Calculation of Beam Deflections by Taylor's Theorem," which appeared in the Appendix.

This book reflects the author's extensive experience in teaching P.E. review courses encompassing almost the entire range of subjects covered in the licensing examinations. The author is indebted to his many students in these courses whose enthusiastic interest in engineering knowledge and its application to our industrial society has proved most stimulating and rewarding.

Max Kurtz

RECOMMENDATIONS TO THE READER

The suggestions that follow are intended to make the study of this book as fruitful as possible and to increase the likelihood that the reader will pass the professional examinations.

1. In studying structural engineering, the most important habit we must develop is that of visualizing, in grossly exaggerated form, the manner in which the given structure or member deforms under the imposed load system. On the basis of this dynamic visualization, we readily perceive the basic nature of the stresses in the structure or the relationship among the reactions.

For example, with reference to Fig. 3-12, we visualize the bar pivoting about the support at A as the load is applied at C, and we thus discern that each hanger elongates by an amount directly proportional to its distance from A. This relationship concerning deformation leads to the basic relationship concerning the reactions. Similarly, with reference to Art. 4-1, we visualize a cylindrical vessel being filled with fluid and expanding transversely and longitudinally in response to the outward thrust of the fluid. We thus conclude that there are tensile stresses in the cylinder shell.

2. In solving many examination problems, considerable benefit accrues if we record the given data in neat tabular form and then enter the calculated values in this table as they emerge. This tabular arrangement has three important advantages. It instantly reveals what gaps still exist in our accumulated data at any given point in the solution, it enables us to locate readily whatever value we must apply in our next calculation, and it contributes to clear thinking through the orderliness of the arrangement. Tables 18-1 and 18-2 in this text serve to illustrate the benefit of solution by tabular arrangement.

3. The study of a subject is rendered far more meaningful and stimulating if we fully understand the reason why that subject is studied. An awareness of the underlying motive enriches our study of the subject and vests it with far greater significance. Therefore, we should strive to understand the motive that underlies our study of each topic.

Let us consider some of these motives. In Chap. 2, we ascribe to an area such properties as statical moment and moment of inertia simply because it has been found that the stresses in a body can conveniently be expressed and analyzed in terms of these properties. Similarly, in Art. 3-3 we investigate the stresses on all planes through a given line in a body mainly because the stresses that we first calculate are not necessarily the critical stresses. In Art. 10-8, we construct influence lines for bridge trusses, not because these lines are of intrinsic significance, but simply because they serve as useful guides in analyzing a truss.

4. Engineers must be flexible in their perception of reality and capable of inverting their viewpoint when it is advantageous to do so. To take an illustration in mechanical engineering, a cam is designed by inverting the motions and considering that the cam remains stationary while the plane in which the follower oscillates revolves about the cam. This inversion of motion is applied because it simplifies design of the cam. Similarly, in structural engineering it is sometimes convenient to invert the relationship between loads and reactions, viewing a load as a reaction and vice versa. Example 7-6 serves as an illustration.

5. Every engineering equation should be tested to determine whether it is rational, i.e., whether it yields the results we anticipate

on the basis of a priori evidence. A rigorous analysis of this type enables us to detect any errors that may have crept into the derivation or recording of the equation and it contributes to a vastly broader understanding.

As an illustration, consider Eq. (7-6), which embodies the theorem of three moments for a continuous beam. This equation gives different expressions for bending moment due to concentrated loads in the left and right spans. Assume that the beam is revolved in a horizontal plane, causing the left and right spans to be interchanged. Performing the calculations, we find that Eq. (7-6) yields identical results for the two positions of the beam, and this consistency tends to substantiate the equation.

Individuals who follow these guidelines will discover that they have considerably sharpened their engineering skills.

ABBREVIATIONS

In general, the identical abbreviation is used for both the singular and plural form.

bhp—brake horsepower
BOD—biochemical oxygen
    demand
cfs—cubic feet per second
cm—centimeter
cu—cubic
fps—feet per second
ft—foot
gal—gallon
gm—gram
gpd—gallons per day
gpm—gallons per minute
in.—inch
kip—1000 pounds
klf—kips per linear foot
ksf—kips per square foot
ksi—kips per square inch
lb—pound

mg—milligram
mgd—million gallons per day
mg/l—milligrams per liter
min—minute
mm—millimeter
pcf—pounds per cubic foot
plf—pounds per linear foot
pli—pounds per linear inch
ppm—parts per million
psf—pounds per square foot
psi—pounds per square inch
rpm—revolutions per minute
sec—second
sq—square
yd—yard
°C—degrees Celsius (formerly
    Centigrade)
°F—degrees Fahrenheit

STRUCTURAL ENGINEERING
FOR PROFESSIONAL ENGINEERS'
EXAMINATIONS

1

ANALYSIS OF FORCE SYSTEMS

1-1 Definitions. A *force* is the tendency of one body to influence the motion of another. It has magnitude and direction.

A *body* force is one that is transmitted to a body through space. Thus, forces of gravitation and magnetism are body forces. If the area through which a body force is transmitted is very small in relation to the area of space under consideration, the force is considered to be transmitted across a straight line, termed its *line of action*. The force is then said to be *concentrated*.

A *surface* force is one that is transmitted to a body by direct contact. If the surface of contact is relatively small, the force is considered to be applied at a point, termed its *point of application*. A force applied at a point is also a concentrated force. A line that contains the point of application and has the same direction as the force is considered to be the line of action of the surface force. A surface force that is distributed across a relatively wide area is said to be *distributed*.

In the material that follows, it is to be understood that all forces are concentrated if nothing is stated to the contrary. The direction of a concentrated force is specified by expressing the angle that its

1

line of action makes with some reference axis, and the sense of the force (northeastward, southeastward, etc.).

A group of forces acting on a body constitute a *force system*. The system is *collinear* if the forces have a common line of action, the system is *coplanar* if the lines of action of all forces lie in one plane, and the system is *concurrent* if all lines of action intersect at a common point. A single force may be regarded as a special type of force system. Two force systems are said to be *equivalent* to one another if they would produce an identical effect on the motion of a given body when applied individually.

In the United States, the basic unit of force is the pound (lb). However, for relatively large forces, it is more convenient to adopt a 1000-lb force as the unit. This force is termed a *kip*, the term being a contraction of kilopound. Thus, a force of 7 kips is 7000 lb.

1-2   Characteristics of Forces and Force Systems.   Every force has three characteristics: magnitude, direction, and position of its line of action. It is convenient to represent a force graphically by means of a vector. This is a line segment that is parallel to the line of action and has a length proportional to the magnitude of the force, with an arrowhead to indicate the sense. Thus, the vector in Fig. 1-1 represents a force of 400 lb that is applied to the body at A, is directed upward to the right in the plane of the drawing, and makes an angle of 30° with the x axis.

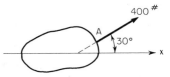

FIG. 1-1. Graphical representation of a force.

If the vector also discloses the point of application of the force, as it does in Fig. 1-1, the vector is said to be *localized*. On the other hand, if the vector discloses only the magnitude and direction of the force, it is said to be *free*.

According to the principle of transmissibility, the *external* behavior of a body under a surface force is determined by the magnitude and line of action of that force, but not by its point of application. As an illustration, consider Fig. 1-2, where body A transmits

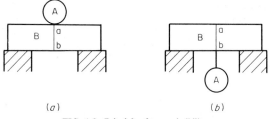

FIG. 1-2. Principle of transmissibility.

its weight (earth pull) to body B, and points a and b lie on a vertical straight line through B. The weight of body A is a force having a vertical line of action. Therefore, with respect to the forces that B transmits to the supporting walls, it is immaterial whether A bears on B at point a, as shown in Fig. 1-2a, or is suspended from B at point b, as shown in Fig. 1-2b. (However, the point at which this force is applied does affect the *internal* behavior of B.)

Every force has a mating force that is numerically equal to and collinear with the given force but oppositely sensed. For example, with reference to Fig. 1-3, consider that bodies A and B are in contact, as shown in Fig. 1-3a. Now consider that A transmits to B the force P shown in Fig. 1-3b. Body B exerts on A the force Q shown in Fig. 1-3c, where P and Q are numerically equal and collinear but oppositely sensed.

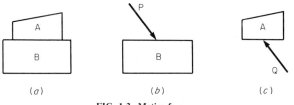

FIG. 1-3. Mating forces.

A force has been defined as the tendency of one body to influence the motion of another. However, the motion of a body consists of two elements, translation and rotation, and it is therefore necessary to appraise the influence of the force on both elements.

The influence of a force on the translation of a body depends on the magnitude and direction of the force. It is independent of the position of its line of action.

The influence of a force on the rotation of a body about a given axis is termed the *moment* or *torque* of the force with respect to that axis. The perpendicular distance from the axis of rotation to the line of action of the force is called the *lever arm* of the force. The axis of rotation is generally understood to be perpendicular to the plane of the drawing, and the point at which this axis intersects the plane of the drawing is called the *moment center*.

The body shown in Fig. 1-4 is connected to its support with a frictionless pin, and the applied force P influences the rotation of the body about the center of pin. Let M denote the moment and d the lever arm. Then

$$M = Pd \qquad (1\text{-}1)$$

In describing a moment, an algebraic sign is used to designate its sense (clockwise or counterclockwise).

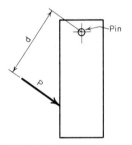

FIG. 1-4. Moment of a force.

Two forces that are equal in magnitude, have parallel but distinct lines of action, and are oppositely sensed constitute a *couple*. A couple is shown in Fig. 1-5, where forces P and Q are equal in mag-

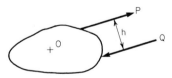

FIG. 1-5. Effect of a couple.

nitude. Let h denote the perpendicular distance between their lines of action. This force system has no effect on the translation of the body, but it does influence its rotation. By selecting an arbitrary moment center O, it is found that the numerical value of the moment of this force system is

$$M = Ph \qquad (1\text{-}2)$$

Thus, the moment of a couple depends solely on the magnitude of the forces and the perpendicular distance between their lines of action. It is independent of the moment center, the specific direction of the forces, and the specific points of application.

When a couple acts on a given body, interest usually centers about the moment of the couple rather than the magnitude of the forces composing the couple. Therefore, the couple is described merely by expressing its moment. Thus, the statement "A clockwise moment of 3000 ft-lb acts on body A" in reality means that a couple or group of couples acts on this body, producing a moment of the specified magnitude. Moreover, if a couple is said to be acting *at a point*, this point usually lies midway between the two points at which the forces composing the couple are applied to the body. The presence of a couple is generally indicated by means of a semicircular arrow, the arrowhead corresponding to the sense of the moment.

1-3   Transformation of Force System.   As stated in Art. 1-1, two force systems are equivalent to one another if they would produce an identical effect on the motion of a given body when applied individually. In analyzing the *external* effects of a force system, it is often convenient to transform the given system to an equivalent system having certain assigned properties. We shall study several such transformations.

In Fig. 1-6a, x and y are rectangular coordinate axes. Assume that the given force system consists of a single force F represented by the vector ab, and that this system is to be transformed to an equivalent system consisting of forces F_x and F_y parallel to the x and y axes, respectively. By projecting ab onto the coordinate axes, we obtain the vectors a_xb_x and a_yb_y that represent F_x and F_y, respectively.

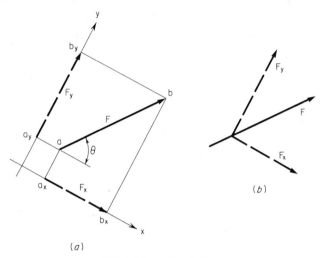

FIG. 1-6. Resolution of a force.

This type of transformation is called the *resolution* of the given force F, and the forces F_x and F_y in the equivalent system are known as the x and y *components* of F, respectively. The magnitude and sense of F_x and F_y determine the influence of F on the motion of a body in their respective directions. In accordance with the principle of transmissibility referred to in Art. 1-2, force F can be resolved into its components at any point whatever along its line of action, as illustrated in Fig. 1-6b.

The sense of each component of F is designated by use of algebraic signs. The component is considered to be positive if its sense agrees with the sense of the corresponding axis, and negative if the reverse is true. Thus, in Fig. 1-6a, both F_x and F_y are positive. Let θ denote the angle between the line of action of F and the x axis. Then

$$F_x = F \cos \theta \qquad\qquad F_y = F \sin \theta \qquad\qquad (1\text{-}3)$$

We shall now consider the reverse situation, in which the components of a force are known and it is necessary to determine the force itself. In Fig. 1-7, x and y are rectangular coordinate axes. The following values are given:

$$F_x = -4000 \text{ lb} \qquad\qquad F_y = 2800 \text{ lb}$$

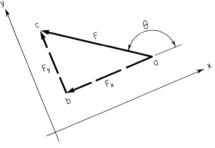

FIG. 1-7

The force F having these components can be found graphically by the following construction: In Fig. 1-7, draw vector ab to represent F_x and vector bc to represent F_y. Now draw vector ac, which represents F.

Analytically, the force F can be found by these calculations:

$$F = \sqrt{4000^2 + 2800^2} = 4883 \text{ lb}$$

$$\tan \theta = \frac{2800}{-4000} = -0.7000 \qquad \theta = 180° - 35° = 145°$$

Now assume that the force system shown in Fig. 1-8 is to be transformed to an equivalent system that meets certain specifications. Establish an arbitrary set of rectangular coordinate axes,

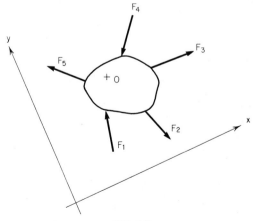

FIG. 1-8

x and y, and consider that all forces in the two systems are resolved into their x and y components. Select an arbitrary moment center O, and consider a moment positive if it tends to rotate the body in a clockwise direction. Let

ΣF_x = algebraic sum of x components of system
ΣF_y = algebraic sum of y components of system
ΣM_O = algebraic sum of moments of system about O

Let the second subscripts 1 and 2 refer to the given force system and its equivalent system, respectively. Since the equivalent system has the same effect on the motion of the body as does the given system, it follows that

$$\Sigma F_{x,2} = \Sigma F_{x,1} \qquad \Sigma F_{y,2} = \Sigma F_{y,1}$$
$$\Sigma M_{O,2} = \Sigma M_{O,1} \qquad (1\text{-}4)$$

In most instances, forces in a coplanar system are resolved into their horizontal and vertical components. Let F_H and F_V denote, respectively, the horizontal and vertical components of a force F. Equations (1-4) assume the following form:

$$\Sigma F_{H,2} = \Sigma F_{H,1} \qquad \Sigma F_{V,2} = \Sigma F_{V,1}$$
$$\Sigma M_{O,2} = \Sigma M_{O,1} \qquad (1\text{-}4a)$$

In this text, we shall adopt the following sign convention, unless otherwise indicated: A horizontal force is positive if it is directed to the right. A vertical force is positive if it is directed upward. A moment is positive if it tends to rotate the body in a clockwise direction.

If the given force system consists solely of a couple, an equivalent system also consists solely of a couple, and the two couples have an identical value of moment as given by Eq. (1-2).

Assume that the force system in Fig. 1-9a is to be transformed to an equivalent system by displacing the force P from A to B while maintaining its original direction, as shown in Fig. 1-9b. Let d denote the perpendicular distance from B to the original line of action of P. By taking moments about B, it is found that the equivalent system consists of the following: the force P applied at B, and a couple or group of couples having a clockwise moment equal to Pd.

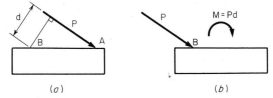

FIG. 1-9. **Transformation of single force to equivalent force and couple.**

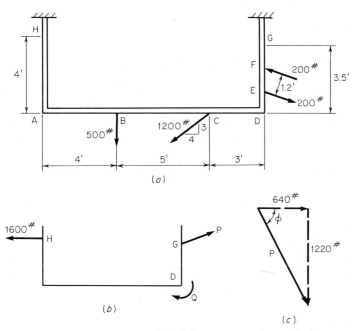

FIG. 1-10

Example 1-1. The U-shaped member in Fig. 1-10a sustains the four forces indicated, the forces at E and F being parallel to one another. This force system is to be replaced with an equivalent system consisting of the following: a horizontal force of 1600 lb to the left applied at H, an inclined force applied at G, and a couple at D. Find the magnitude and direction of the force at G and the moment of the couple. Verify the results.

Solution. In addition to the four forces that are shown, the bar sustains resisting forces at the supports. However, these resisting forces are not part of the force system under consideration.

Refer to Fig. 1-10*b*. Let *P* denote the inclined force at *G* and *Q* denote the moment of the couple at *D*. Assume that these are sensed as shown.

The forces at *E* and *F* in the given system constitute a couple having a moment of $-200 \times 1.2 = -240$ ft-lb. These forces may be excluded in calculating the algebraic sum of components. The action line of the force at *C* forms a 3-4-5 right triangle with horizontal and vertical lines. For this force,

$$F_H = -\tfrac{4}{5} \times 1200 = -960 \text{ lb}$$
$$F_V = -\tfrac{3}{5} \times 1200 = -720 \text{ lb}$$

For the given force system,

$$\Sigma F_H = -960 \text{ lb} \qquad \Sigma F_V = -500 - 720 = -1220 \text{ lb}$$

Select *G* as the moment center. (In recording the value of a moment, a subscript is used to identify the moment center.) To find the moment of the force at *C*, replace this force with its horizontal and vertical components, applied at *C*. Taking the moments of the force at *B*, the force at *C*, and the couple, in that order, we obtain the following:

$$\Sigma M_G = -500 \times 8 + 960 \times 3.5 - 720 \times 3 - 240 = -3040 \text{ ft-lb}$$

When the foregoing values are applied to the equivalent system shown in Fig. 1-10*b*, the following equations result:

$$\Sigma F_H = -1600 + P_H = -960 \qquad \therefore P_H = 640 \text{ lb}$$
$$\Sigma F_V = P_V = -1220 \text{ lb}$$
$$\Sigma M_G = -1600 \times 0.5 + Q = -3040 \qquad \therefore Q = -2240 \text{ ft-lb}$$

The negative value of *Q* signifies that the couple at *D* has a counterclockwise moment. Refer to Fig. 1-10*c*, which shows the magnitude and direction of *P*.

$$P = \sqrt{640^2 + 1220^2} = 1378 \text{ lb}$$

$$\tan \phi = \frac{1220}{640} = 1.9063 \qquad \phi = 62°19'$$

The foregoing results can be verified in this manner: Select *C* as moment center. For the given system,

$$\Sigma M_C = -500 \times 5 - 240 = -2740 \text{ ft-lb}$$

For the equivalent system,

$$\Sigma M_C = -1600 \times 4 + 640 \times 3.5 + 1220 \times 3 - 2240$$
$$= -2740 \text{ ft-lb} \qquad \text{OK}$$
$$P_H = 1378 \cos 62°19' = 640 \text{ lb} \qquad \text{OK}$$

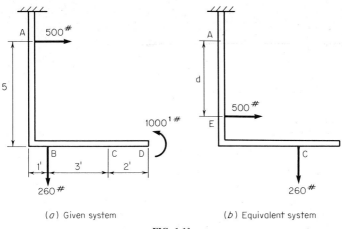

(a) Given system   (b) Equivalent system

FIG. 1-11

Example 1-2. The force system in Fig. 1-11a is to be replaced with an equivalent system by displacing the vertical force of 260 lb from B to C and eliminating the couple of 1000 ft-lb at D. How far must the horizontal force of 500 lb be displaced?

Solution. Refer to Fig. 1-11b, where d denotes the displacement of the horizontal force. Select A as the moment center. For the given system,

$$\Sigma M_A = 260 \times 1 - 1000 = -740 \text{ ft-lb}$$

For the equivalent system,

$$\Sigma M_A = -500d + 260 \times 4 = -740 \qquad d = \textbf{3.56 ft}$$

1-4 Resultant of Force System. Associated with every force system is an infinite set of equivalent systems. The *simplest* equivalent system is called the *resultant* of the system, and the process of replacing a given force system with its resultant is known as the *composition of forces.*

There are two types of resultants, as follows:

1. If the given system has a nonzero value of ΣF_H, ΣF_V, or both, the resultant is a single force. Equations (1-4a) assume the following form:

$$R_H = \Sigma F_H \qquad R_V = \Sigma F_V$$
$$M_{O,R} = \Sigma M_O \tag{1-5}$$

where the terms on the left pertain to the resultant and the terms on the right pertain to the given system. The forces of the given system are components of the resultant.

2. If the given system has a zero value of both ΣF_H and ΣF_V but a nonzero value of ΣM_O, the resultant may be considered to be a couple having a moment equal to ΣM_O.

Example 1-3. With reference to the force system in Fig. 1-12a, find the magnitude, direction, and point of application of the resultant if it is to be applied to the vertical leg of the frame.

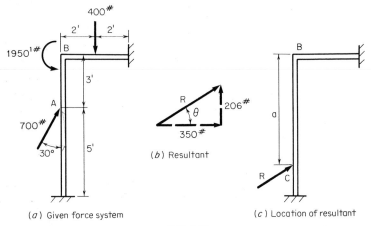

(a) Given force system   *(c)* Location of resultant

(b) Resultant

FIG. 1-12

Solution. For the force at A,

$$F_H = 700 \sin 30° = 350 \text{ lb} \qquad F_V = 700 \cos 30° = 606 \text{ lb}$$

Then
$$R_H = \Sigma F_H = 350 \text{ lb}$$
$$R_V = \Sigma F_V = 606 - 400 = 206 \text{ lb}$$

Consider the force at A to be resolved into its components at A.

$$M_{B,R} = \Sigma M_B = -1950 + 400 \times 2 - 350 \times 3 = -2200 \text{ ft-lb}$$

Refer to Fig. 1-12b.

$$R = \sqrt{350^2 + 206^2} = \mathbf{406\ lb}$$

$$\tan \theta = \frac{206}{350} = 0.5886 \qquad \theta = \mathbf{30°29'}$$

Refer to Fig. 1-12c, which shows R applied to the frame at C. Resolving R into its components at C and applying the value of $M_{B,R}$ calculated above, we obtain

$$M_{B,R} = -350a = -2200 \qquad a = \mathbf{6.29\ ft}$$

The magnitude and direction of the resultant of a force system can be found by a simple graphical procedure. Refer to Fig. 1-13a, which shows a system of five forces acting on a body. In Fig. 1-13b, draw the chain of free vectors, ab, bc, cd, de, and ef, to represent the indicated forces. Now draw vector af from the initial point a to the terminal point f of the chain, placing the arrowhead at f. The force represented by vector af has horizontal and vertical components equal, respectively, to ΣF_H and ΣF_V of the given force system. Therefore, the force represented by af is the resultant of the system.

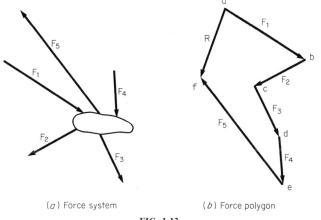

(*a*) Force system   (*b*) Force polygon

FIG. 1-13

The vector diagram in Fig. 1-13b is known as a *force polygon*. In drawing the force polygon of a force system for the purpose of obtaining the magnitude and direction of the resultant, the forces may be taken in any sequence whatever. Each sequence yields a unique force polygon, but the initial and terminal points of the vector chain occupy the same positions relative to one another in all these polygons.

1-5 Conditions of Equilibrium. A force system that has a zero resultant is referred to as a *balanced* system, and the body on which it acts is said to remain *in equilibrium.* In accordance with Eqs. (1-5), a balanced force system has the following characteristics:

$$\Sigma F_H = 0 \qquad \Sigma F_V = 0 \qquad \Sigma M_O = 0 \qquad (1\text{-}6)$$

The foregoing are referred to as the *equations of equilibrium.*

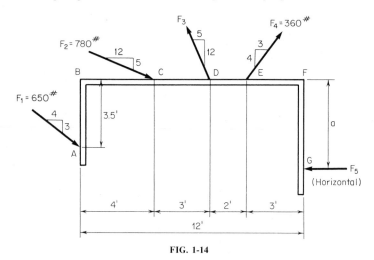

FIG. 1-14

Example 1-4. The body in Fig. 1-14 remains in equilibrium under the force system shown. Determine the magnitude of F_3 and the magnitude and location of F_5.

Solution. The action lines of the inclined forces form 3-4-5 and 5-12-13 right triangles with horizontal and vertical lines. The horizontal and vertical components of these forces, in absolute value, are as follows:

$$F_{1,H} = \tfrac{4}{5} \times 650 = 520 \text{ lb} \qquad F_{1,V} = \tfrac{3}{5} \times 650 = 390 \text{ lb}$$
$$F_{2,H} = \tfrac{12}{13} \times 780 = 720 \text{ lb} \qquad F_{2,V} = \tfrac{5}{13} \times 780 = 300 \text{ lb}$$
$$F_{4,H} = \tfrac{3}{5} \times 360 = 216 \text{ lb} \qquad F_{4,V} = \tfrac{4}{5} \times 360 = 288 \text{ lb}$$

With respect to the system,

$$\Sigma F_H = 520 + 720 - F_{3,H} + 216 - F_5 = 0 \qquad (a)$$
$$\Sigma F_V = -390 - 300 + F_{3,V} + 288 = 0 \qquad (b)$$

From Eq. (b), $F_{3,V} = 402 \text{ lb}$

Then $F_{3,H} = \tfrac{5}{12} \times 402 = 167.5 \text{ lb}$
$$F_3 = \tfrac{13}{12} \times 402 = \textbf{435.5 lb}$$

From Eq. (*a*), $F_5 = 1288.5 \text{ lb}$

Select F as moment center.

$$\Sigma M_F = -520 \times 3.5 - 390 \times 12 - 300 \times 8 + 402 \times 5 + 288 \times 3 + 1288.5a$$
$$= 0$$

Solving, $a = 4.68 \text{ ft}$

Refer again to Fig. 1-13*b*, where the magnitude and direction of the resultant of a force system were found graphically. If the system has a zero resultant, the chain of free vectors closes upon itself. Therefore, in the case of a balanced force system, the vector chain and the force polygon are identical.

In accordance with Eqs. (1-6), there is a set of three simultaneous equations associated with every balanced coplanar force system. However, if the force system is concurrent, it can be demonstrated that one equation in the set, selected at random, can be obtained by combining multiples of the remaining two equations. Consequently, the set consists of only two *independent* equations.

1-6   Classification of Forces.   In structural analysis, we usually deal with a body that is connected to supports in order to hold the body at rest. As forces are applied to the body, resisting forces are induced at the supports. The forces that tend to move the body are known as active forces or *loads*; the forces that tend to resist motion of the body are known as passive forces or *reactions*. If the body does remain at rest, the complete system of forces acting on the body, consisting of the loads and reactions, constitutes a balanced force system.

The loads acting on a structural member may be divided into two categories: *dead* loads, which are permanent and constant, and *live* loads, which are transient and variable. For example, with respect to a beam supporting the roof of a building, the weight of the beam and the weight of the roofing material are dead loads; the weight of the snow on the roof is a live load.

The reaction offered by a support can assume the form of a single force, a couple, a combination of a single force and couple, or some equivalent system. For example, in many instances the reaction is a distributed force of varying intensity that is equivalent to a single force and couple.

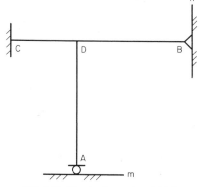

FIG. 1-15. Structural supports and joints.

1-7   Structural Supports and Joints. There are three basic types of structural supports, as illustrated in Fig. 1-15. At A, the structure bears on its support by means of a smooth (i.e., frictionless) roller. Since friction is absent, the support offers resistance to displacement of end A solely in the direction normal to the m axis, and therefore the reaction of the support is normal to this axis.

At B, the structure is connected to its support with a smooth pin, thereby preventing linear displacement but not rotation of end B. The reaction of this support is a single force, and since B is fully constrained against displacement, this force has a component both normal and parallel to the n axis. The support at B is said to be *simple* or *hinged*.

At C, the structure is tied to its support in a manner that prevents both linear displacement and rotation of end C. The reaction of the support is equivalent to a system consisting of a single force that prevents linear displacement and a couple that clamps the end in position and thereby prevents rotation. This type of support is said to be *fixed*.

A structural joint, such as that at D in Fig. 1-15, is termed *flexible* if each member at the joint is free to rotate relative to any other, and *rigid* if relative rotation is prevented. At a rigid joint, the members of the frame are constrained to undergo an identical rotation.

Where nothing is stated to the contrary, it is to be understood that all supports are hinged and all joints are flexible, except in the case of reinforced-concrete construction.

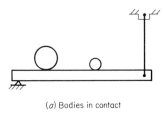

(*a*) Bodies in contact

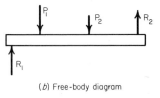

(*b*) Free-body diagram

FIG. 1-16

1-8   Free-Body Diagrams and Internal Forces.   Assume   that   a given body receives surface forces from contiguous bodies. The force analysis can be simplified by isolating the body from its surroundings and exhibiting solely the body and its force system, which is represented by vectors. The body is then described as *free*, and the resulting diagram is called a *free-body diagram*. The procedure is illustrated in Fig. 1-16.

The forces induced within a body, which are termed *internal* forces, can be unearthed by dividing the body into segments and treating each segment as an independent body that remains in equilibrium. In this manner, the internal forces are transformed to external forces and thus become amenable to analysis.

As an illustration, consider that the rectangular bar in Fig. 1-17*a* is in equilibrium under the three forces indicated. Conceive the bar to be severed along the *m* axis; the upper segment of the bar is shown as a free body in Fig. 1-17*b*. By applying the equations of equilibrium, we deduce that the internal forces distributed along the *m* axis have the following equivalent force system: a force F_i, applied at the center, that is equal and parallel to F_1 but oppositely sensed, and a couple having a counterclockwise moment $M = F_1 h$, where h is the perpendicular distance between the lines of action of F_1 and F_i. The internal forces we have investigated are the forces that the lower part of the bar exerts on the upper part.

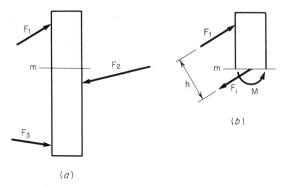

(a)

(b)

FIG. 1-17. Cutting of member to find internal forces.

Example 1-5.   A load of 1000 lb is supported by two flexible cables in the manner shown in Fig. 1-18a. Determine the tension in each cable.

Solution.   Let T_A and T_B denote the tensile forces in the cables supported at A and B, respectively, and let the second subscripts H and V refer to the horizontal and vertical components, respectively.

METHOD 1:   Figure 1-18b is the free-body diagram of the pin at C. Applying the first two equations of equilibrium, we obtain the following:

$$\Sigma F_H = -T_{A,H} + T_{B,H} = 0 \qquad (a)$$
$$\Sigma F_V = T_{A,V} + T_{B,V} - 1000 = 0 \qquad (b)$$

But
$$T_{A,V} = T_{A,H} \tan 35° = 0.700 T_{A,H}$$
and
$$T_{B,V} = T_{B,H} \tan 40° = 0.839 T_{B,H}$$

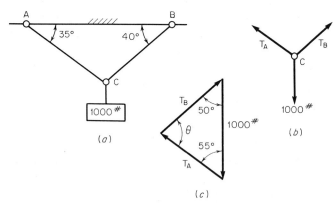

(a)

(b)

(c)

FIG. 1-18

Substituting these values in Eq. (*b*) gives

$$0.700T_{A,H} + 0.839T_{B,H} - 1000 = 0 \qquad (c)$$

Solving Eqs. (*a*) and (*c*) gives

$$T_{A,H} = 650 \text{ lb} \qquad T_{B,H} = 650 \text{ lb}$$

Then $\qquad T_A = T_{A,H} \sec 35° = 650 \times 1.221 = \textbf{794 lb}$

and $\qquad T_B = T_{B,H} \sec 40° = 650 \times 1.305 = \textbf{848 lb}$

METHOD 2: In a situation of this type, it is usually much simpler to find the unknown forces by analyzing the force polygon. Figure 1-18*c* is the force polygon of the balanced force system acting on the pin at *C*. The angles that forces T_A and T_B make with the vertical are recorded in the drawing, and

$$\theta = 180° - (50° + 55°) = 75°$$

By the law of sines,

$$\frac{T_A}{1000} = \frac{\sin 50°}{\sin 75°} = \frac{0.766}{0.966} \qquad T_A = 793 \text{ lb}$$

$$\frac{T_B}{1000} = \frac{\sin 55°}{\sin 75°} = \frac{0.819}{0.966} \qquad T_B = 848 \text{ lb}$$

1-9 Analysis of Statically Determinate Structures. Assume that a structure remains at rest under a coplanar force system, and consider that all forces are resolved into their horizontal and vertical components. In accordance with Art. 1-7, the structure can have three types of reactions: horizontal forces, vertical forces, and couples. The couples are described by means of their moments.

It is possible to develop a set of simultaneous equations for this structure by applying the equations of equilibrium to the structure and to individual parts of the structure. If this set of simultaneous equations enables us to evaluate the reactions, the structure is described as *statically determinate* with respect to its external force system.

In calculating reactions, it is necessary to assume the sense of a force or moment if the sense is not readily apparent. If the calculated value is positive, the force or moment has the sense assumed; if the calculated value is negative, the force or moment has the opposite sense.

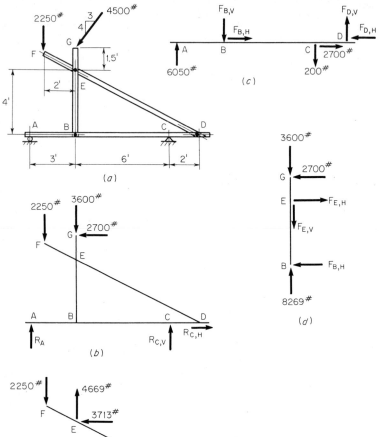

FIG. 1-19

Example 1-6. The frame in Fig. 1-19a rests on a smooth roller at A and is tied to a support at C. The members of the frame have negligible weight. Find the horizontal and vertical components of the reactions and of the forces acting at pins B, D, and E.

Solution. A force will be identified by means of a subscript corresponding to its point of application. For example, F_F denotes the force applied at F. Then

$$F_{G,H} = \tfrac{3}{5} \times 4500 = 2700 \text{ lb} \qquad F_{G,V} = \tfrac{4}{5} \times 4500 = 3600 \text{ lb}$$

Figure 1-19b is the free-body diagram of the frame, the members being represented by straight lines. The thickness of the horizontal member may be disregarded in calculating distances. Since the roller at A is smooth, the reaction at this support is normal to the surface and therefore vertical. Applying the equations of equilibrium, we obtain the following:

$$\Sigma F_H = R_{C,H} - 2700 = 0$$
$$\Sigma F_V = R_A + R_{C,V} - 2250 - 3600 = 0 \qquad (a)$$

or
$$R_A + R_{C,V} - 5850 = 0 \qquad (b)$$
$$\Sigma M_C = 9R_A - 2250 \times 8 - 3600 \times 6 - 2700 \times 5.5 = 0$$

or
$$9R_A - 54{,}450 = 0 \qquad (c)$$

From Eq. (c), $R_A = 6050 \text{ lb}$

From Eq. (a), $R_{C,H} = 2700 \text{ lb}$

From Eq. (b), $R_{C,V} = -200 \text{ lb}$

That $R_{C,V}$ is negative signifies that the true sense of this force is opposite to that assumed, and the force therefore acts downward.

Figure 1-19c is the free-body diagram of the horizontal bar. This drawing shows the magnitude and true sense of every force that is known at this point. Four unknown forces are present, but three of these can be eliminated by taking moments about D.

$$\Sigma M_D = 6050 \times 11 - 8F_{B,V} - 200 \times 2 = 0$$
$$\therefore F_{B,V} = 8269 \text{ lb}$$

The positive result signifies that the assumed sense of this force is correct.

$$\Sigma F_V = 6050 - 200 - 8269 + F_{D,V} = 0$$
$$\therefore F_{D,V} = 2419 \text{ lb}$$

Figure 1-19d is the free-body diagram of the vertical bar. In proceeding from Fig. 1-19c to 1-19d, it is necessary to reverse the sense of each force at B, for this reason: Figure 1-19c shows the forces that the vertical bar is exerting on the horizontal bar at B. Conversely, Fig. 1-19d shows the forces that the horizontal bar is exerting on the vertical bar at B. In Fig. 1-19d,

$$\Sigma F_V = 8269 - F_{E,V} - 3600 = 0 \qquad \therefore \; F_{E,V} = 4669 \text{ lb}$$
$$\Sigma M_B = -2700 \times 5.5 + 4F_{E,H} = 0 \qquad \therefore \; F_{E,H} = 3713 \text{ lb}$$
$$\Sigma F_H = -2700 + 3713 - F_{B,H} = 0 \qquad \therefore \; F_{B,H} = 1013 \text{ lb}$$

Returning to Fig. 1-19c with this value of $F_{B,H}$, we have

$$\Sigma F_H = 1013 + 2700 - F_{D,H} = 0 \qquad \therefore \; F_{D,H} = 3713 \text{ lb}$$

In summary, the forces at the pins are as follows:

Pin	B	D	E
Horizontal force, lb	1013	3713	3713
Vertical force, lb	8269	2419	4669

The foregoing results can be verified by demonstrating that the forces on the inclined bar satisfy the equations of equilibrium. Refer to Fig. 1-19e.

$$\Sigma F_H = -3713 + 3713 = 0 \qquad\qquad\qquad \text{OK}$$
$$\Sigma F_V = -2250 + 4669 - 2419 = 0 \qquad\qquad \text{OK}$$
$$\Sigma M_D = -2250 \times 10 + 4669 \times 8 - 3713 \times 4 = 0 \qquad \text{OK}$$

Example 1-7. Bar AB in Fig. 1-20a weighs 80 lb, and it is connected to the wall by means of a pin at A and a horizontal cable at B. A cylinder of 2-ft diameter and weighing 300 lb is placed between the vertical wall and bar AB. If all contact surfaces are smooth, find the reaction of the wall at A, E, and C. The reaction at A can be expressed in terms of its horizontal and vertical components.

Solution. The thickness of the bar can be disregarded. Refer to Fig. 1-20b. Line AO bisects the angle between the wall and bar AB.

$$OD = \text{radius of cylinder} = 1 \text{ ft}$$
$$AD = 1 \times \cot 18° = 3.08 \text{ ft}$$

Figure 1-20c is the free-body diagram of the cylinder. Since friction is absent, the forces at E and D are normal to the wall and bar, respectively, and therefore radial with respect to the cylinder. Angle θ equals the angle EAD because their sides are mutually perpendicular. From the equations of equilibrium, we have

$$F_{D,V} = 300 \text{ lb}$$

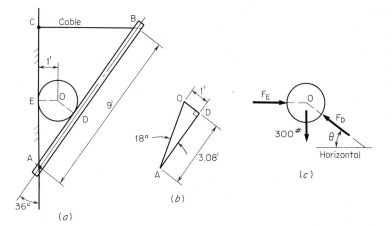

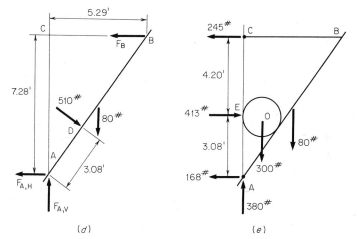

FIG. 1-20

Then
$$F_{D,H} = 300 \cot 36° = 300 \times 1.376 = 413 \text{ lb}$$
$$F_E = F_{D,H} = 413 \text{ lb}$$
$$F_D = 300 \csc 36° = 300 \times 1.701 = 510 \text{ lb}$$

In Fig. 1-20a,

$$AC = 9 \cos 36° = 9 \times 0.809 = 7.28 \text{ ft}$$
$$BC = 9 \sin 36° = 9 \times 0.588 = 5.29 \text{ ft}$$

Figure 1-20d is the free-body diagram of bar AB. The bar is assumed to be homogeneous, and the resultant of its weight is therefore a concentrated force acting at the center of bar. In this drawing,

$$\Sigma M_A = 510 \times 3.08 + 80(\tfrac{1}{2} \times 5.29) - 7.28F_B = 0$$
$$\therefore F_B = 245 \text{ lb}$$

The tension in the cable and the reaction of the wall at C are therefore both 245 lb. Applying the horizontal and vertical components of the force at D, we obtain the following:

$$\Sigma F_H = -F_{A,H} + 413 - 245 = 0 \qquad \therefore F_{A,H} = 168 \text{ lb}$$
$$\Sigma F_V = F_{A,V} - 300 - 80 = 0 \qquad \therefore F_{A,V} = 380 \text{ lb}$$

The reactions of the wall are shown in Fig. 1-20e. The results can be tested by taking the bar, cable, and cylinder as a composite body and testing the equilibrium of this body.

$$\Sigma F_H = -245 + 413 - 168 = 0 \qquad\qquad\qquad \text{OK}$$
$$\Sigma F_V = 380 - 300 - 80 = 0 \qquad\qquad\qquad\qquad \text{OK}$$
$$\Sigma M_C = 168 \times 7.28 - 413 \times 4.20 + 300 \times 1 + 80 \times 2.645 = 0 \qquad \text{OK}$$

1-10   Static Friction.   In Fig. 1-21a, bodies A and B are in contact with one another along a surface that is nominally a plane. Body A is subjected to a force system equivalent to the system shown, where P is normal to the surface of contact of the bodies and Q is parallel thereto. Figure 1-21b is the free-body diagram of A, with the reaction of B on A resolved into its normal and parallel components.

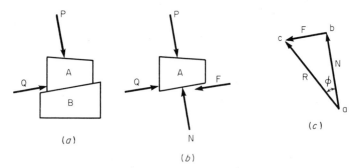

FIG. 1-21. Static friction.

In response to Q, body A tends to slide along B, but B resists this relative displacement by exerting a force F on A parallel to the surface of contact, as shown in Fig. 1-21b. This resisting force arises from the asperities of the abutting faces, and it is termed *static friction*. If A remains stationary relative to B, we have

$$N = P \qquad \text{and} \qquad F = Q$$

Assume that Q is gradually increased until the system has mobilized the maximum potential friction and sliding of A along B therefore impends. The limiting value of F is directly proportional to the normal force P (and therefore to N). The constant of proportionality is called the *coefficient of static friction* and is denoted by μ. Then

$$F_{\max} = \mu N \tag{1-7}$$

The value of μ depends on the materials of which bodies A and B are composed and the degree of roughness of their abutting faces.

The reaction of body B on A is represented by vector ac in Fig. 1-21c. Let ϕ denote the angle between the line of action of the resultant and a line normal to the surface of contact of the bodies. As seen in this drawing, $\phi = \arctan F/N$. Let $\phi_{\max}$ denote the value of ϕ at impending motion; this is known as the *angle of friction*. Then

$$\phi_{\max} = \arctan \mu \tag{1-8}$$

Therefore, if the angle of friction is given, the coefficient of friction can be found by setting $\mu = \tan \phi_{\max}$.

In Fig. 1-22a, body A of weight W rests on an inclined plane having an angle of inclination θ. In Fig. 1-22b, vector ac represents the

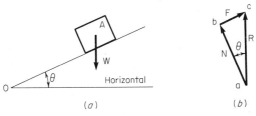

FIG. 1-22. Angle of repose.

reaction R of the plane on A, and vectors ab and bc represent the components of R normal and parallel to the plane, respectively. Body A tends to slide down the plane in response to the component of W parallel to the plane, but this tendency is opposed by the frictional force F of the plane. If A remains stationary, the reaction R is vertical and equal to W, and angle bac in Fig. 1-22b equals the angle of inclination θ.

Now assume that the plane rotates about O in a counterclockwise direction until A is on the verge of sliding. Let θ_{max} denote the value of θ at impending motion; this is known as the *angle of repose* of A with respect to the plane. By setting $F = F_{max} = \mu N$, we obtain from Fig. 1-22b

$$\theta_{max} = \arctan \mu \qquad (1\text{-}9)$$

Thus, the angle of repose equals the angle of friction.

Example 1-8. In Fig. 1-23a, bodies A and B weigh 160 lb and 280 lb, respectively. The bodies are placed in the positions shown and then released. The coefficient of friction between B and the horizontal support is 0.15, and

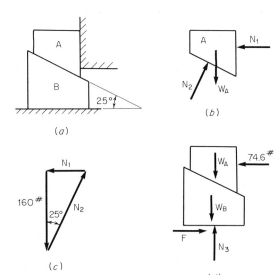

FIG. 1-23

all other contact surfaces are smooth. Determine whether B slides to the left or remains at rest.

Solution. Where the dimensions of the bodies are omitted, it is understood that the only possible form of motion is translation. Consequently, a body remains in equilibrium if it satisfies the two equations $\Sigma F_H = 0$ and $\Sigma F_V = 0$.

In response to the component of its weight parallel to the surface of contact, body A tends to slide to the right *relative to B*. However, since the wall constrains A against an absolute horizontal movement to the right, body B tends to slide to the left.

We shall assume that the system remains at rest, calculate the amount of friction required to maintain this condition, and then compare this value with the maximum potential friction. Figure 1-23b is the free-body diagram of A, and Fig. 1-23c is the corresponding force polygon. Then

$$N_1 = 160 \tan 25° = 160 \times 0.466 = 74.6 \text{ lb}$$

Figure 1-23d is the free-body diagram of A and B taken as a composite body. Then

$$N_3 = 160 + 280 = 440 \text{ lb} \qquad F = 74.6 \text{ lb}$$

The maximum potential friction is

$$F_{\max} = \mu N_3 = 0.15 \times 440 = 66 \text{ lb}$$

Since the frictional resistance that can be mobilized is less than that required to hold B at rest, this body slides to the left.

Example 1-9. In Fig. 1-24a, bodies A and B weigh 220 lb and 350 lb, respectively. The coefficients of friction are as follows: between A and the

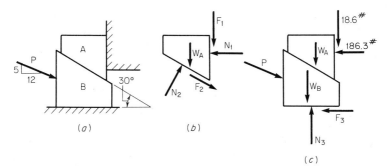

FIG. 1-24

wall, 0.01; between A and B, 0.14; between B and the horizontal surface, 0.20. Find the minimum value of P that is required to elevate A.

Solution.   Body A is elevated by a movement of B to the right.

Figure 1-24b is the free-body diagram of A. Since A tends to move upward, the frictional resistance of the wall is downward. Similarly, since A tends to move to the left *relative to* B, the frictional resistance of B is to the right. If the system remains at rest, we have the following:

$$\Sigma F_H = -N_1 + N_2 \sin 30° + F_2 \cos 30° = 0 \qquad (a)$$
$$\Sigma F_V = -F_1 + N_2 \cos 30° - F_2 \sin 30° - 220 = 0 \qquad (b)$$

At impending motion,

$$F_1 = 0.10 N_1 \qquad \text{and} \qquad F_2 = 0.14 N_2$$

With the appropriate substitutions, Eqs. (a) and (b) assume the following forms, respectively:

$$-N_1 + 0.500 N_2 + 0.866 \times 0.14 N_2 = 0$$
$$-0.10 N_1 + 0.866 N_2 - 0.500 \times 0.14 N_2 - 220 = 0$$

or
$$-N_1 + 0.621 N_2 = 0 \qquad (c)$$
and
$$-0.10 N_1 + 0.796 N_2 = 220 \qquad (d)$$

Solving Eqs. (c) and (d) for N_1 gives

$$N_1 = 186.3 \text{ lb} \qquad \therefore F_1 = 18.6 \text{ lb}$$

Figure 1-24c is the free-body diagram of bodies A and B taken as a composite body. The line of action of P forms a 5-12-13 right triangle with horizontal and vertical lines. At impending motion,

$$\Sigma F_H = P_H - F_3 - 186.3 = 0$$
or
$$\tfrac{12}{13} P - 0.20 N_3 = 186.3 \qquad (e)$$
$$\Sigma F_V = -P_V + N_3 - (220 + 350 + 18.6) = 0$$
or
$$-\tfrac{5}{13} P + N_3 = 588.6 \qquad (f)$$

Solving Eqs. (e) and (f),

$$P = \mathbf{359 \text{ lb}}$$

1-11   Belt Friction.   In Fig. 1-25a, belt AB is wrapped about the stationary cylinder along the arc CD, and forces T_L and T_S are applied at the ends of the belt, T_L being the larger and T_S the smaller of the forces. Arc CD is referred to as the *arc of contact*, and the central angle θ that this arc subtends is called the *angle of contact*.

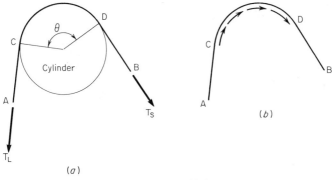

FIG. 1-25. Belt friction.

The tendency of the belt to slip over the cylinder as a result of the unequal forces is resisted by friction between the cylinder and belt along the arc of contact, as shown in Fig. 1-25b. As a result of the friction, the tensile force in the belt diminishes from T_L at C to T_S at D, the rate of decrease in tension being directly proportional to the magnitude of the tension at the given point. Consider that the difference between T_L and T_S is increased until slipping impends. At this state,

$$T_L = T_S e^{\mu\theta} \tag{1-10}$$

where e = base of natural logarithms = 2.718 (to four significant figures)

μ = coefficient of friction between cylinder and belt

θ = angle of contact, radians

To convert an angular measure from degrees to radians, we have the following:

$$\pi \text{ radians} = 180° \qquad \therefore 1° = 0.01745 \text{ radian}$$

Example 1-10. In Fig. 1-26a, body A weighs 300 lb. It is held in position by a belt that passes over a fixed drum, the belt being subjected to a horizontal force P at its end. The coefficient of friction is 0.18 between A and the inclined plane and 0.25 between the belt and drum. Find the value of P when A is at impending motion (a) up the plane; (b) down the plane.

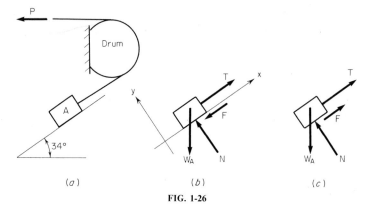

FIG. 1-26

Solution

$$\theta = 180° - 34° = 146°$$
or
$$\theta = 146 \times 0.01745 = 2.548 \text{ radians}$$
$$\mu\theta = 0.25 \times 2.548 = 0.637 \qquad e^{\mu\theta} = 2.718^{0.637} = 1.891$$

PART *a*: Figure 1-26*b* is the free-body diagram of *A*. Establish x and y reference axes parallel and normal to the plane, respectively, as shown.

$$\Sigma F_y = N - 300 \cos 34° = 0$$
or
$$N - 0.829 \times 300 = 0$$
$$N = 249 \text{ lb} \qquad F = 0.18 \times 249 = 44.8 \text{ lb}$$
$$\Sigma F_x = T - 44.8 - 300 \sin 34° = 0$$
or
$$T - 44.8 - 0.559 \times 300 = 0 \qquad T = 212.5 \text{ lb}$$

Force T is the force that the belt exerts on A and, conversely, the force that A exerts on the belt. Since A is on the verge of sliding up the plane, $P > T$. By Eq. (1-10),

$$P = Te^{\mu\theta} = 212.5 \times 1.891 = \mathbf{402 \text{ lb}}$$

PART *b*: Figure 1-26*c* is the free-body diagram of *A*. As before, $N = 249$ lb and $F = 44.8$ lb.

$$\Sigma F_x = T + 44.8 - 300 \sin 34° = 0 \qquad T = 122.9 \text{ lb}$$

In the present instance, $T > P$. Then

$$P = \frac{122.9}{1.891} = \mathbf{65.0 \text{ lb}}$$

PROBLEMS

1-1 The force system in Fig. 1-27 is to be replaced with an equivalent system consisting of a horizontal force applied at *b* and a vertical force applied to the horizontal leg. Find the magnitude of the vertical force and its point of application.

ANS. 700 lb; 3.14 ft to right of *d*

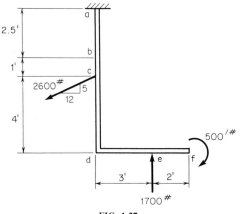

FIG. 1-27

1-2 In Fig. 1-28, a vertical force *P* applied at *b* induces tensile forces of 500 lb and 660 lb in cables *ab* and *bc*, respectively. Find the value of *P* and θ.

ANS. 748 lb; 49°

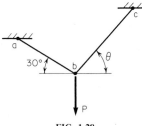

FIG. 1-28

1-3 With reference to the frame in Fig. 1-29, find the tension *T* in the cable and the horizontal and vertical components of the reaction R_A at *A*.

ANS. $T = 8206$ lb; $R_{A,H} = 1907$ lb to left; $R_{A,V} = 203$ lb downward

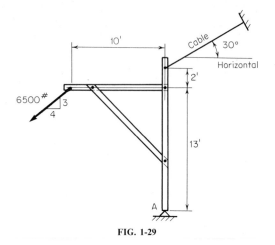

FIG. 1-29

1-4 In Fig. 1-30, cylinders A and B are placed on the inclined planes and released. When equilibrium is achieved, their line of centers O_1O_2 is found to make an angle of $16°$ with the horizontal, as shown. If A weighs 360 lb and all surfaces are smooth, what is the weight of B?   ANS. 299 lb

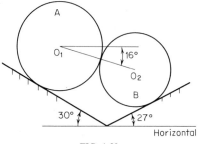

FIG. 1-30

1-5 The block in Fig. 1-31a weighs 150 lb, and the coefficient of static friction between the block and the inclined plane is 0.28. If the block is on the verge of sliding up the plane under the 120-lb force, what is the value of angle θ?

ANS. $19°01'$ or $72°17'$

HINT: Draw the force polygon, as shown in Fig. 1-31b. Draw the diagonal ac, and equate angle dca to the angle of friction. Then find angle acb by the law of sines. Since angle acb can be either acute or obtuse, there are two possible values of θ.

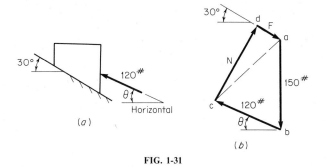

FIG. 1-31

1-6 Bodies *A* and *B* in Fig. 1-32 weigh 500 lb and 100 lb, respectively. They are supported by a cord of negligible weight that is wrapped about a stationary drum. If the coefficient of static friction between *A* and *B* is 0.25, what is the minimum value of the coefficient of static friction between the cord and drum if the system is to remain at rest?   ANS. 0.364

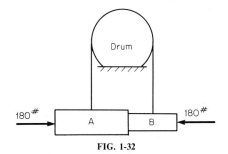

FIG. 1-32

2

GEOMETRICAL PROPERTIES
OF AN AREA

2-1 Objective. Assume that a balanced force system is applied to a structural member composed of a single material and that we wish to evaluate the stresses in the member acting along some plane normal to the plane of the force system. These stresses are governed by the following: the characteristics of the force system, the area of the member as projected onto that plane, and the manner in which this area is distributed.

Since certain expressions pertaining to an area arise repeatedly in stress analysis, it is advantageous to assign a nomenclature to these expressions and to analyze the properties thus defined.

2-2 Statical Moment. With reference to Fig. 2-1a, consider that figure *abcd* is resolved into elements, as indicated. Let dA denote the area of an element and y denote the ordinate of the center of the element. The product $y\,dA$ is termed the *statical moment* of the element with respect to the x axis. The statical moment of the total area of *abcd* is defined as the algebraic sum of the statical moments

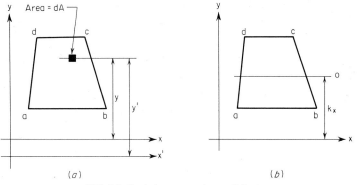

FIG. 2-1. Statical moment and centroidal axis.

of its elements. Then

$$Q_x = \int y \, dA \qquad (2\text{-}1)$$

where Q_x = statical moment of total area with respect to x axis.

Similarly, the statical moment of the total area with respect to the x' axis is

$$Q_{x'} = \int y' \, dA$$

The distance y' is positive or negative according to whether the element lies above or below the x' axis. Since an area is always regarded as positive, the algebraic sign of $y' \, dA$ agrees with that of y'. In general, therefore, the statical moment of the area with respect to a given axis can be positive, negative, or zero. The particular axis about which the statical moment is zero is termed the *centroidal axis* of the area.

In Fig. 2-1b, let o denote the centroidal axis of the area of $abcd$, and let k_x denote the distance that this axis lies above the x axis. From the definition of statical moment, we have the following:

$$Q_o = \int (y - k_x) \, dA = \int y \, dA - k_x \int dA = Q_x - Ak_x$$

where A denotes the total area. Since $Q_o = 0$, we obtain

$$Q_x = Ak_x \qquad (2\text{-}2a)$$

and

$$k_x = \frac{Q_x}{A} \qquad (2\text{-}2b)$$

In many instances, the statical moment of an area with respect to a given axis is known or can readily be determined, and it is necessary to locate the centroidal axis parallel to the given axis. This can be done by applying Eq. (2-2b) to find the distance k between these axes.

Assume that the given area contains an axis of symmetry parallel to the x axis. In taking the statical moment of the area with respect to the axis of symmetry, we find that every absolute value of $y \, dA$ occurs in pairs: one positive, the other negative. Therefore, an axis of symmetry is a centroidal axis.

Now consider that the x and y axes are rotated to the new positions x' and y', respectively, as shown in Fig. 2-2. By calculating $Q_{x'}$ and applying Eq. (2-2), we locate a new centroidal axis o' that is parallel to the x' axis. Manifestly, an area has an infinite number of centroidal axes, each corresponding to an assigned direction of the x axis. It can be demonstrated that all centroidal axes are concurrent (i.e., they intersect at a common point), and this point of concurrence G is called the *centroid* of the area. To locate the centroid, it is merely necessary to locate two centroidal axes and find their point of intersection.

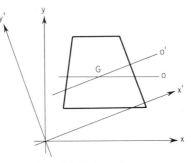

FIG. 2-2. Centroid of area.

Since statical moment is a sum, the statical moment of a composite area is the aggregate of the statical moments of its parts. In Fig. 2-3, consider that the two areas shown, of magnitudes A_1 and A_2, are combined to form a composite area. Let o_1 and o_2 denote the centroidal axes of the given areas and o_c denote the centroidal axis of the composite area. By taking moments about any convenient

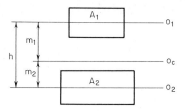

FIG. 2-3. Centroidal axis of composite area.

axis to locate o_c, we obtain the following results, in absolute value:

$$m_1 = h\,\frac{A_2}{A_1 + A_2} \qquad \text{and} \qquad m_2 = h\,\frac{A_1}{A_1 + A_2} \tag{2-3}$$

The geometrical properties of various shapes are recorded in standard reference books. The reader should commit to memory the following properties of a triangle and trapezoid: In Fig. 2-4a,

$$k = \frac{d}{3} \tag{2-4}$$

$$Q_{MN} = \frac{ad^2}{6} \tag{2-5}$$

In Fig. 2-4b,

$$k = \frac{d}{3}\,\frac{a + 2b}{a + b} \tag{2-6}$$

$$Q_{MN} = \frac{d^2}{6}\,(a + 2b) \tag{2-7}$$

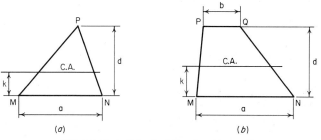

FIG. 2-4. Centroidal axis of triangle and trapezoid.

An alternative expression for statical moment is *first moment.* If the unit of length is the in., the unit of statical moment is the in.³.

Example 2-1. Area *abcdefg* in Fig. 2-5 is symmetrical with respect to a vertical line through *e.* Locate the centroid of this area.

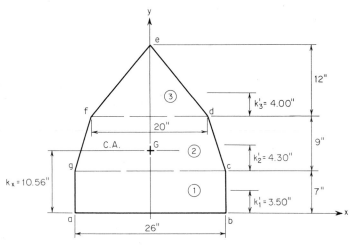

FIG. 2-5. Centroid of composite area.

Solution. Let G denote the centroid. Establish rectangular coordinate axes in the positions indicated. Since y is an axis of symmetry, it is the vertical centroidal axis. To locate the horizontal centroidal axis, proceed as follows:

Divide area *abcdefg* into segments by drawing the dashed lines, and label the segments in the manner shown. Let k' denote the distance from the base of a segment to its horizontal centroidal axis. Then

$$k'_1 = \tfrac{1}{2} \times 7 = 3.50 \text{ in.}$$

$$k'_2 = \frac{9}{3}\left(\frac{26 + 2 \times 20}{26 + 20}\right) = 4.30 \text{ in.} \qquad k'_3 = \tfrac{1}{3} \times 12 = 4.00 \text{ in.}$$

Calculate the statical moment of area *abcdefg* with respect to the x axis by finding the statical moments of its segments. The calculations are recorded in the accompanying table, where the superscript ′ signifies that the quantity pertains to a segmental area as distinguished from the total area.

Segment	Area, sq in.	$\times$	k'_x, in.	$=$	Q'_x, in.^{3}
1	$7 \times 26 = 182$		3.50		637
2	$\frac{1}{2}(26 + 20)9 = 207$		11.30		2339
3	$\frac{1}{2} \times 20 \times 12 = 120$		20.00		2400
Total	509				5376

By Eq. (2-2),

$$k_x = \frac{Q_x}{A} = \frac{5376}{509} = 10.56 \text{ in.}$$

The centroid G has the position indicated.

2-3   Resultant of Distributed Force.   It will be convenient at this point to discuss an important application of the concept of centroidal axis. Assume that a body is subjected to a distributed force acting in one plane. The *pressure* is the magnitude of the force acting on a unit length of the body. If the pressure varies, the distributed force can be represented by means of a pressure diagram.

In Art. 1-2, we defined the moment of a force about an axis of rotation with reference to a *concentrated* force. The moment of a *distributed* force can be found by dividing the pressure diagram into elemental strips, taking the moments of these strips with respect to the given axis, and then aggregating these moments. Thus, the moment of a distributed force about a given axis equals the statical moment of its pressure diagram about that axis.

If the pressure diagram of the distributed force has some standard shape, the moment of the force can most readily be found by replacing the distributed force with its resultant, this being an equivalent *concentrated* force. According to the definition, the resultant has the same magnitude and direction as the distributed force, and its moment about any axis is equal to that of the distributed force. It follows that the resultant has a magnitude equal to the area of the pressure diagram and that its line of action lies at the centroidal axis of this diagram.

As an illustration, assume that bar AB in Fig. 2-6 is subjected to a distributed force such that the pressure varies linearly from 500 plf

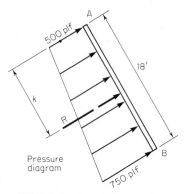

FIG. 2-6. Resultant of distributed force.

(lb per linear ft) at A to 750 plf at B. The resultant R has the following magnitude and location:

$$R = \tfrac{1}{2}(500 + 750)18 = 11{,}250 \text{ lb}$$

$$k = \frac{18}{3}\left(\frac{500 + 2 \times 750}{500 + 750}\right) = 9.6 \text{ ft}$$

Therefore, the moment of the distributed force with respect to A is

$$M_A = 11{,}250 \times 9.6 = 108{,}000 \text{ ft-lb}$$

2-4   Moment of Inertia.   With reference to Fig. 2-7, consider again that figure *abcd* is resolved into elements, as indicated. Continuing the previous notation, let dA denote the area of an element and y

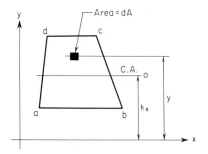

FIG. 2-7. Moment of inertia.

denote the ordinate of the center of the element. Let

$$I_x = \int y^2 \, dA \tag{2-8}$$

The quantity I_x is termed the *moment of inertia* of the total area with respect to the x axis.

By extending the foregoing definition, we can calculate the moment of inertia of the total area with respect to any axis parallel to the x axis by replacing y with the distance from the center of the element to the axis under consideration.

Let o denote the centroidal axis parallel to the x axis and I_o denote the moment of inertia of the area with respect to that axis. Then

$$I_o = \int (y - k_x)^2 \, dA = \int y^2 \, dA - 2k_x \int y \, dA + k_x^2 \int dA$$
$$= I_x - 2k_x Q_x + A k_x^2$$

By replacing Q_x with $A k_x$ and rearranging terms, we obtain

$$I_x = I_o + A k_x^2 \tag{2-9}$$

This relationship is known as the *transfer formula* for moment of inertia.

The moment of inertia of an area is always positive, and Eq. (2-9) discloses that it has a minimum value with respect to the centroidal axis. If the unit of length is the in., the unit of moment of inertia is the in.^{4}.

By analogy with Eq. (2-2a), it is desirable to express the moment of inertia of an area in terms of the total area A, and we accordingly write the following:

$$I_x = A r_x^2 \tag{2-10}$$

The distance r_x introduced for this purpose is called the *radius of gyration* of the area with respect to the x axis. Solving Eq. (2-10) for r_x gives

$$r_x = \sqrt{\frac{I_x}{A}} \tag{2-10a}$$

If Eq. (2-9) is recast in terms of radius of gyration, it becomes transformed to

$$A r_x^2 = A r_o^2 + A k_x^2$$

where r_o is the radius of gyration of the area with respect to the centroidal axis. Then

$$r_x{}^2 = r_o{}^2 + k_x{}^2 \tag{2-11}$$

The rectangle in Fig. 2-8 has the following moment of inertia with respect to the centroidal axis indicated:

$$I_o = \frac{bd^3}{12} \tag{2-12}$$

The triangle in Fig. 2-4a has the following moment of inertia with respect to the centroidal axis indicated:

$$I_o = \frac{ad^3}{36} \tag{2-13}$$

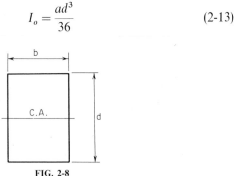

FIG. 2-8

Moment of inertia, like statical moment, is a sum, and therefore the moment of inertia of a composite area is the aggregate of the moments of inertia of its segments. Equation (2-9) can therefore be written in this form with respect to a composite area:

$$I_x = \Sigma I_{o'} + \Sigma A' k_x'^2 \tag{2-9a}$$

where A' = area of segment
   $I_{o'}$ = moment of inertia of segmental area with respect to centroidal axis of segment
   k_x' = distance from centroidal axis of segment to x axis

Example 2-2.   Compute the moment of inertia of the area in Fig. 2-9 with respect to its horizontal centroidal axis.

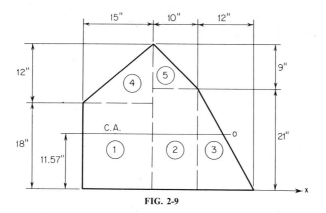

FIG. 2-9

Solution. It is first necessary to locate this axis. Place the x axis at the base, and divide the area into rectangles and triangles by drawing the dashed lines. The statical moment of the area about the x axis is calculated in the accompanying table.

Segment	Area, sq in.	$\times$	k'_x, in.	$=$	Q'_x, in.^{3}
1	270		9.0		2430
2	210		10.5		2205
3	126		7.0		882
4	90		22.0		1980
5	45		24.0		1080
Total	741				8577

Then
$$k_x = \frac{8577}{741} = 11.57 \text{ in.}$$

The moment of inertia of the area with respect to this centroidal axis will now be found by applying Eq. (2-9a). Replacing x with o, we obtain

$$I_o = \Sigma I_{o'} + \Sigma A' k'^2_o$$

where the subscripts o and o' refer to the centroidal axes of the total area and of a segmental area, respectively. The value of k'_o is found by deducting 11.57 from the value of k'_x recorded above; only absolute values are needed. Refer to the accompanying table.

Segment	$I_{o'}$, in.^{4}		k_o', in.	$A'k_o'^2$, in.^{4}
1	$\frac{1}{12} \times 15 \times 18^3 =$	7290	2.57	1783
2	$\frac{1}{12} \times 10 \times 21^3 =$	7718	1.07	240
3	$\frac{1}{36} \times 12 \times 21^3 =$	3087	4.57	2632
4	$\frac{1}{36} \times 15 \times 12^3 =$	720	10.43	9790
5	$\frac{1}{36} \times 10 \times 9^3 =$	203	12.43	6953
Total		19,018		21,398

Then $\qquad\qquad I_o = 19{,}018 + 21{,}398 = \textbf{40,416 in.}^4$

2-5   Polar Moment of Inertia; Product of Inertia.   With reference to Fig. 2-10a, consider that the area is resolved into elements, as indicated. Let dA denote the area of an element, x and y its rectangular coordinates, and ρ its radius vector. Let

$$J_Q = \int \rho^2 \, dA \qquad (2\text{-}14)$$

The quantity J_Q is known as the *polar moment of inertia* of the area with respect to an axis that passes through the pole Q and is perpendicular to the plane of the area.

Equation (2-14) can be recast in this form:

$$J_Q = \int \rho^2 \, dA = \int (y^2 + x^2) \, dA = \int y^2 \, dA + \int x^2 \, dA$$
$$\therefore J_Q = I_x + I_y \qquad (2\text{-}15)$$

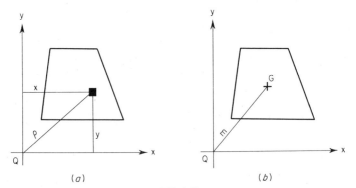

(a) $\qquad\qquad\qquad\qquad$ (b)

FIG. 2-10

It therefore follows that if the x and y axes are rotated into new positions the sum of the moments of inertia of the area with respect to these axes remains constant, since J_Q remains constant.

In Fig. 2-10b, let G denote the centroid of the area, m the radius vector of the centroid, and J_G the polar moment of inertia of the area with respect to an axis through G. Then

$$J_Q = J_G + Am^2 \qquad (2\text{-}16)$$

Now let

$$P_{xy} = \int xy \, dA \qquad (2\text{-}17)$$

The quantity P_{xy} is known as the *product of inertia* of the area with respect to the x and y axes.

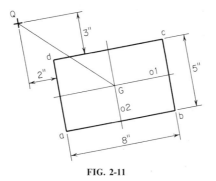

FIG. 2-11

Example 2-3. With reference to Fig. 2-11, find the polar moment of inertia of the rectangle *abcd* with respect to an axis through Q.

Solution. Draw the centroidal axes *o1* and *o2* parallel to the sides, and call G the centroid.

$$I_{o1} = \tfrac{1}{12} \times 8 \times 5^3 = 83.3 \text{ in.}^4$$
$$I_{o2} = \tfrac{1}{12} \times 5 \times 8^3 = 213.3 \text{ in.}^4$$
$$J_G = I_{o1} + I_{o2} = 296.6 \text{ in.}^4 \qquad A = 8 \times 5 = 40 \text{ sq in.}$$
$$(QG)^2 = (3 + 2.5)^2 + (2 + 4)^2 = 66.25$$
$$J_Q = J_G + A(QG)^2 = 296.6 + 40 \times 66.25 = \mathbf{2946.6 \text{ in.}^4}$$

2-6 Statical Moment of Partial Area.

In Fig. 2-12, o is the horizontal centroidal axis of the area shown, and u is an arbitrary horizontal axis. Let $Q_{o,\text{above}}$ and $Q_{o,\text{below}}$ denote the statical moment

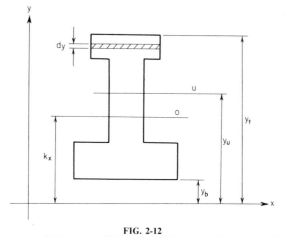

FIG. 2-12

with respect to o of that part of the area that lies above u and below u, respectively. Thus,

$$Q_{o,\text{above}} = \int_{y_u}^{y_t} (y - k_x)\, dA$$

$$Q_{o,\text{below}} = \int_{y_b}^{y_u} (y - k_x)\, dA \qquad (2\text{-}18)$$

From the definition of centroidal axis, it follows that

$$Q_{o,\text{above}} + Q_{o,\text{below}} = 0$$
$$\therefore Q_{o,\text{below}} = -Q_{o,\text{above}} \qquad (2\text{-}19)$$

Consider that the entire area is divided into elemental horizontal strips of thickness dy, as indicated. The statical moment of a strip with respect to o is positive or negative according to whether the strip lies above or below o, respectively. The quantity $Q_{o,\text{above}}$ is the algebraic sum of the statical moments of all strips that lie above axis u.

Now consider that u is displaced vertically downward a distance dy, causing $Q_{o,\text{above}}$ to change by the statical moment of the strip through which u has been displaced. This change in $Q_{o,\text{above}}$ is positive or negative according to whether the original position of u was above or below axis o, respectively. It therefore follows that $Q_{o,\text{above}}$ has its maximum value when u coincides with o.

PROBLEM

2-1. With reference to Fig. 2-13, locate the centroid G of the area shown, and compute the moment of inertia of this area with respect to the x and o axes.

ANS. $k_x = 4.69$ in., $k_y = 5.46$ in.; $I_x = 2439$ in.^{4}, $I_o = 2439 - 78 \times 4.69^2 = 723$ in.^{4}

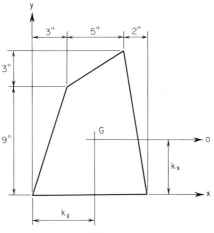

FIG. 2-13

3

ANALYSIS OF STRESS
AND STRAIN

3-1 Objective. When the external forces acting on a member have been evaluated, the next step in structural analysis consists of evaluating the stresses and strains that the force system engenders within the member. This chapter accordingly will develop the basic relationships pertaining to stress and strain and then apply those relationships to typical situations that arise in practice.

3-2 Definitions. Figure 3-1*a* and *b* shows a straight and a curved structural member, respectively. Consider that a plane is passed through each member, normal to the direction of the member at that location. The surface of intersection of this transverse plane and the member is known as the *cross* (or *transverse*) *section* of the member. A line that runs across the length of the member and contains the centroid of every cross section is known as the *longitudinal axis* of the member. From these definitions, it follows that the longitudinal axis is normal to every cross section. If the cross section is uniform across the entire length of the member, the member is said to be *prismatic*. The member shown in Fig. 3-1*c* has a variable cross section and is therefore nonprismatic. A member composed

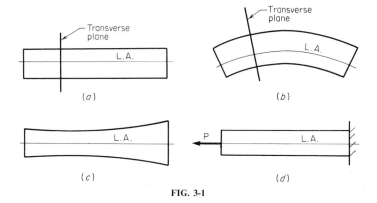

FIG. 3-1

of a single material is *homogeneous,* and one composed of more than one material is *composite.*

Consider a force to be acting on a member. If the line of action of the force passes through the centroid of a given cross section, the force is *concentric* with respect to that section; if the line of action is normal to the cross section, the force is *normal* with respect to that section. If the member is straight and the line of action of the force coincides with the longitudinal axis, the force is described as *axial.* Thus, the force P in Fig. 3-1d is axial. It follows that a force that is axial with respect to the member is both concentric and normal with respect to every cross section. A normal force is either *tensile* or *compressive,* depending on the character of the concomitant strain.

The force acting on a unit area of a member, either at the exterior surface or within the interior of the member, is termed the *stress.* Consider that a stress is resolved into components normal and parallel to the area on which it acts. The normal component is referred to as a *normal, direct,* or *axial* stress; the parallel component is referred to as a *shearing* stress. A stress acting on the exterior surface of a member is often referred to as a *pressure.* The usual unit of stress is psi (lb per sq in.).

Conceive the straight members in Fig. 3-2a and c to be severed along the transverse plane MM. The parts below this plane are shown as free bodies in Fig. 3-2b and d, respectively. Since each body is in equilibrium, the distributed internal force at MM has a

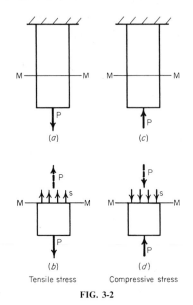

(a)

(c)

(b)

(d)

Tensile stress   Compressive stress

FIG. 3-2

resultant that is equal to and collinear with the applied force *P*. If this force is axial, the internal force is assumed to be uniformly distributed across the section. Let *A* denote the area of the cross section and *s* the normal stress at *MM*. Then

$$s = \frac{P}{A} \tag{3-1}$$

Example 3-1.   A steel hanger having a cross-sectional area of 2.09 sq in. carries an axial load of 26 kips. Compute the tensile stress in the member.

Solution

$$s = \frac{P}{A} = \frac{26,000 \text{ lb}}{2.09 \text{ sq in.}} = \mathbf{12{,}400 \text{ psi}}$$

The prismatic member in Fig. 3-3a is supported at face *ABCD* and is subjected to a vertical load *P* on the other face, the action line of *P* passing through the centroid of the section. Conceive the

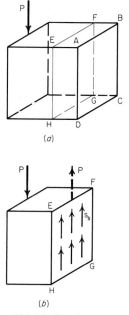

FIG. 3-3. Shearing stress.

member to be severed at the section $EFGH$. The part to the left of this section is shown in Fig. 3-3b. Since the member is in equilibrium, the distributed internal force on $EFGH$ has an equivalent system consisting of a vertical concentrated force equal to P and passing through the centroid of the section, and a couple. If the moment of the couple is negligible, the shearing stress on $EFGH$ is assumed to be uniform.

3-3 Stresses on Collinear Planes; Mohr's Circle. Figure 3-4a shows a body that remains in equilibrium under a coplanar force system, and A is a line through the body normal to the plane of the force system. Consider that a plane Q is passed through A, and let f and s_s denote the normal and shearing stress, respectively, acting on this plane at A. The values of f and s_s constitute the *state of stress* at A with respect to this plane.

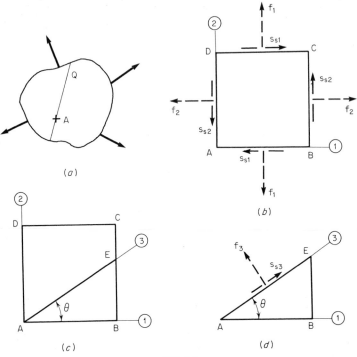

FIG. 3-4

Now consider that horizontal and vertical planes are passed through A and through an adjacent line to form the infinitesimally small block $ABCD$. Figure 3-4b shows this block and the stresses on its faces, the stresses being represented by dashed vectors as a means of distinguishing them from forces. The horizontal and vertical planes through A are labeled 1 and 2, respectively, and the stresses on the faces of the block are assigned subscripts to match. Since the block is infinitesimally small, corresponding stresses on opposite faces may be considered to be equal. We shall adopt the following sign convention with respect to stresses:

A normal stress is positive if it is tensile. A shearing stress is positive if it tends to rotate the body on which it acts in a clockwise direction.

In accordance with this convention, the stresses f_1, f_2, and s_{s1} in Fig. 3-4b are all positive, and s_{s2} is negative. The shearing stresses on opposite faces constitute a couple, and by applying the equilibrium equation $\Sigma M = 0$, we find that $s_{s2} = -s_{s1}$. We thus arrive at the following relationship:

THEOREM 3-1: The shearing stresses on two mutually perpendicular planes are equal in absolute value but differ in algebraic sign.

To ascertain whether the body in Fig. 3-4a can support the imposed force system, it is necessary to perform a stress analysis and to identify the critical stresses that exist in the member. As a starting point, assume that the stresses shown in Fig. 3-4b are known; our next step is to determine the state of stress at A with respect to other planes. For this purpose, pass an arbitrary plane through A in Fig. 3-4c, and call this plane 3. The location of plane 3 will be specified by expressing the value of angle θ between this plane and plane 1, which is taken as the reference plane. Angle θ will be considered acute, and the sign convention is as follows: The angle is positive if we go from plane 1 to plane 3 in a counterclockwise direction.

Let f_3 and s_{s3} denote, respectively, the normal and shearing stresses on plane 3, as shown in Fig. 3-4d. By drawing a free-body diagram of ABE, resolving the forces on this block into components normal and parallel to plane 3, and applying the equations of equilibrium, we obtain the following results:

$$f_3 = \frac{f_1 + f_2}{2} + \frac{f_1 - f_2}{2} \cos 2\theta - s_{s1} \sin 2\theta \qquad (3\text{-}2a)$$

$$s_{s3} = \frac{f_1 - f_2}{2} \sin 2\theta + s_{s1} \cos 2\theta \qquad (3\text{-}2b)$$

Since the block $ABCD$ lies in the interior of the body, the arbitrary plane 3 can have any position whatever, and therefore angle θ can range from $-90°$ to $90°$.

Equations (3-2) can be represented graphically by means of the diagram shown in Fig. 3-5, which is known as *Mohr's circle of stress*. The construction is as follows:

1. Selecting an origin O, draw a horizontal axis to represent values of f and a vertical axis to represent values of s_s.

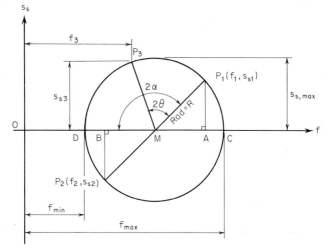

FIG. 3-5. Mohr's circle of stress.

2. Plot points P_1 and P_2 having coordinates equal to the stresses on planes 1 and 2, respectively. (In this drawing, it has been assumed that f_1, f_2, and s_{s1} are positive and that $f_1 > f_2$.) Then

$$OA = f_1 \qquad OB = f_2 \qquad AP_1 = s_{s1} \qquad BP_2 = s_{s2} = -s_{s1}$$

3. Draw line P_1P_2, intersecting the f axis at M. Since $AP_1 = -BP_2$, M is the midpoint of P_1P_2 and of BA. Then

$$OM = \frac{f_1 + f_2}{2} \qquad MA = \frac{f_1 - f_2}{2}$$

4. Draw a circle having M as center and MP_1 as radius.
5. Draw the radius MP_3 making an angle of 2θ with MP_1. If θ is positive, we go from MP_1 to MP_3 in the counterclockwise direction.

It can readily be demonstrated that the coordinates of P_3 equal the stresses on plane 3, as given by Eqs. (3-2).

Since plane 3 was arbitrarily selected, it follows that the state of stress on every plane through line A in the body is represented by a unique point on Mohr's circle, and conversely that every point on

Mohr's circle represents the state of stress on a unique plane through A. The angle between two radii in Mohr's circle is twice the angle between the corresponding planes in the body.

Figure 3-5 reveals that OM is the arithmetical mean of the normal stresses on any pair of mutually perpendicular planes. Therefore, $2(OM)$ is the sum of these stresses. We thus arrive at the following conclusion with respect to the set of planes through line A:

THEOREM 3-2: The sum of the normal stresses on two mutually perpendicular planes is constant.

The maximum shearing stress on a plane through line A in the body equals the radius of Mohr's circle, and the maximum and minimum normal stresses on planes through A equal OC and OD, respectively, in Fig. 3-5. Let R denote the radius. Then

$$R = \sqrt{(MA)^2 + (AP_1)^2}$$

or
$$R = \sqrt{\left(\frac{f_1 - f_2}{2}\right)^2 + s_{s1}{}^2} \tag{3-3a}$$

$$s_{s,\max} = R \qquad\qquad s_{s,\min} = -R \tag{3-3b}$$

$$f_{\max} = \frac{f_1 + f_2}{2} + R \qquad f_{\min} = \frac{f_1 + f_2}{2} - R \tag{3-3c}$$

Let α denote the angle between plane 1 and the plane of $f_{\min}$. From Fig. 3-5, we have the following:

$$\tan 2\alpha = -\frac{s_{s1}}{(f_1 - f_2)/2} \tag{3-3d}$$

If s_{s1} is positive, 2α lies between 0 and $180°$; if s_{s1} is negative, 2α lies between 0 and $-180°$. The planes on which $f_{\max}$ and $f_{\min}$ occur are sometimes referred to as the *major* and *minor* planes, respectively.

A plane on which $s_s = 0$ and $f \neq 0$ is termed a *principal plane*, and the normal stress on this plane is termed a *principal stress*. Figure 3-5 reveals that when $s_s = 0$ the normal stress has either its maximum or minimum value. We therefore arrive at the following conclusions with respect to the set of planes through A:

THEOREM 3-3: There are two principal planes, and they are mutually perpendicular.

THEOREM 3-4: A principal stress is either the maximum or minimum normal stress.

THEOREM 3-5: The planes of maximum and minimum shearing stress make angles of 45° with the principal planes.

Example 3-2. A rectangular block *ABCD* lying in the interior of a body has the stresses indicated in Fig. 3-6*a*.

a. Determine the state of stress on plane 3, and verify the answer.

b. Find the principal stresses and the maximum and minimum shearing stresses, and identify the planes on which these stresses occur.

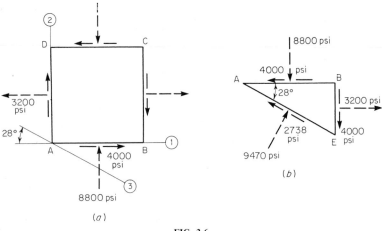

(a)

(b)

FIG. 3-6

Solution. We shall first evaluate the stresses and locate the principal planes by applying Eqs. (3-2) and (3-3). Then we shall construct Mohr's circle.

$$f_1 = -8800 \text{ psi} \qquad f_2 = 3200 \text{ psi} \qquad s_{s1} = -4000 \text{ psi}$$

$$\frac{f_1 + f_2}{2} = -2800 \text{ psi} \qquad \frac{f_1 - f_2}{2} = -6000 \text{ psi}$$

PART *a*. With respect to plane 3,

$$\theta = -28° \qquad 2\theta = -56°$$
$$\sin 2\theta = -0.829 \qquad \cos 2\theta = 0.559$$

By Eqs. (3-2),

$$f_3 = -2800 + (-6000)0.559 - (-4000)(-0.829)$$
$$= -9470 \text{ psi (compression)}$$
$$s_{s3} = -6000(-0.829) + (-4000)0.559 = 2738 \text{ psi}$$

To verify these values, refer to Fig. 3-6b, which shows the stresses on the triangular block *ABE*. Set $AE = 1$ in.

$$AB = \cos 28° = 0.883 \qquad BE = \sin 28° = 0.470$$

Consider the block to be 1 in. thick. Then

$$\Sigma F_H = -4000 \times 0.883 + 3200 \times 0.470 + 9470 \times 0.470$$
$$- 2738 \times 0.883 \simeq 0 \qquad \text{OK}$$
$$\Sigma F_V = -8800 \times 0.883 - 4000 \times 0.470 + 9470 \times 0.883$$
$$+ 2738 \times 0.470 \simeq 0 \qquad \text{OK}$$

PART *b*. By Eqs. (3-3),

$$R = \sqrt{6000^2 + 4000^2} = 7211 \text{ psi}$$
$$s_{s,\max} = 7211 \text{ psi} \qquad s_{s,\min} = -7211 \text{ psi}$$
$$f_{\max} = -2800 + 7211 = 4411 \text{ psi (tension)}$$
$$f_{\min} = -2800 - 7211 = -10{,}011 \text{ psi (compression)}$$
$$\tan 2\alpha = -\frac{-4000}{-6000} = -0.6667$$

The arctan of 0.6667 is 33°42′. Since s_{s1} is negative, we set

$$2\alpha = -33°42′ \qquad \text{and} \qquad \alpha = -16°51′$$

Mohr's circle of stress is constructed in Fig. 3-7a, and the location of the specified planes is shown in Fig. 3-7b.

A plane on which $f = 0$ and $s_s \neq 0$ is said to be in a state of *pure shear*. This state exists only if Mohr's circle intersects the s_s axis, as it does in Fig. 3-7a. On the basis of this drawing, we can state the following with respect to the set of planes through line *A*:

THEOREM 3-6a: A state of pure shear exists only if $f_{\max}$ is tensile and $f_{\min}$ is compressive.

THEOREM 3-6b: If a state of pure shear does exist, it occurs on two planes that are symmetrically located with respect to the principal

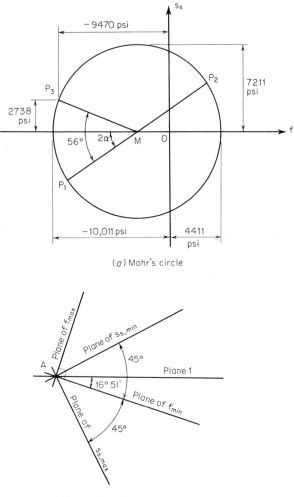

(*a*) Mohr's circle

(*b*) Location of significant planes

FIG. 3-7

planes. The shearing stresses on these planes are equal in absolute value but differ in algebraic sign.

With reference to the body in Fig. 3-6, the location of the planes of pure shear and the stresses on these planes can be found by constructing Fig. 3-8.

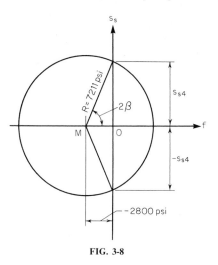

FIG. 3-8

3-4   Deformation of a Stressed Body.   The configuration of a body is modified when a balanced force system is applied. If the body reverts completely to its original shape upon removal of the force system, it is described as *elastic*; if, on the other hand, the body retains a permanent set, it is described as *plastic*. Structural materials generally exhibit elasticity below a particular stress and plasticity beyond that stress, the point of demarcation being referred to as the *elastic limit*.

There are two compelling reasons for studying the manner in which a body deforms under a force system. First, deformation is often a criterion in structural design because excessive deformation of a member will have detrimental effects. Second, as we shall find in Art. 3-5, a detailed study of the deformation of a structure is often an essential aspect of the stress analysis of the structure.

Figure 3-9 shows the deformation of a member resulting from axial forces. Let L denote the original length and ΔL the change in length of the member. If the member is prismatic and homogeneous, the change in length of a unit length is the ratio $\Delta L/L$. This quantity is termed the *axial strain* of the member; it is denoted by ϵ. Then

$$\epsilon = \frac{\Delta L}{L} \tag{3-4}$$

Strain is a dimensionless quantity.

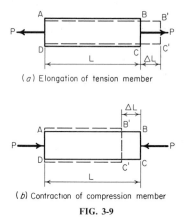

(*a*) Elongation of tension member

(*b*) Contraction of compression member

FIG. 3-9

For most structural materials there is a direct proportionality between the simultaneous values of stress and strain up to a particular value of stress called the *proportional limit*. We shall generally assume that our stresses fall below the proportional limit. The ratio of a given axial stress to its corresponding axial strain is termed the *modulus of elasticity* or *Young's modulus* of the material; it is denoted by *E*. If *s* denotes the axial stress, we have

$$E = \frac{s}{\epsilon} \tag{3-5}$$

The unit of modulus of elasticity coincides with that of stress.
Equation (3-5) can be extended in this manner:

$$E = \frac{s}{\epsilon} = \frac{P/A}{\Delta L/L}$$

Rearranging,
$$\Delta L = \frac{PL}{AE} = \frac{sL}{E} \tag{3-6}$$

Example 3-3. A steel hanger having a cross-sectional area of 2.25 sq in. carries an axial load of 26 kips. If the hanger is 8 ft long and $E = 30,000,000$ psi, what is the displacement of the load?

Solution

$$\Delta L = \frac{PL}{AE} = \frac{26,000 \times 8 \times 12}{2.25 \times 30,000,000} = \textbf{0.037 in.}$$

If the member is nonprismatic or the internal force is not constant, the change in length must be found by summation or integration. As shown in Fig. 3-9, a body under axial forces deforms along the line of action of the forces and along each line normal to the line of action. These orthogonal strains are opposite in character, one being an elongation and the other a contraction. Let ϵ_x and ϵ_y denote the strains in the direction of the axial forces and in a transverse direction, respectively. The absolute value of ϵ_y/ϵ_x is known as *Poisson's ratio*. Let m denote this ratio, and consider an elongation to be positive and a contraction to be negative. Then

$$m = -\frac{\epsilon_y}{\epsilon_x} \tag{3-7}$$

Example 3-4. A prismatic member that is 20 ft long sustains a tensile force that causes it to elongate 0.18 in. If $m = 0.24$, what is the transverse strain?

Solution

$$\epsilon_x = \frac{\Delta L}{L} = \frac{0.18}{20 \times 12} = 0.00075$$

$$\epsilon_y = -m\epsilon_x = -0.24 \times 0.00075 = \mathbf{-0.00018}$$

When a steel member is stressed in tension beyond the proportional limit, a point is eventually reached at which the member continues to elongate in the absence of any increase in stress. The member is then said to *yield*, and the stress at which yielding occurs is called the *yield-point stress*.

With reference to Fig. 3-10, *ABCD* is a square block having faces in a state of pure shear. The shearing stresses cause the block to

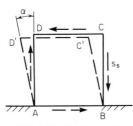

FIG. 3-10. Shearing strain.

assume the shape $ABC'D'$. Let s_s denote the absolute value of the shearing stress on each face and α the angle through which one face rotates relative to another. The *shearing strain* ϵ_s is defined as the tangent of angle α. However, since this angle in reality is extremely small, we may equate the tangent to the angle itself, expressed in radians. Then

$$\epsilon_s = \alpha \tag{3-8}$$

The ratio of shearing stress to shearing strain is called the *modulus of rigidity* of the material; it is denoted by G. Then

$$G = \frac{s_s}{\epsilon_s} \tag{3-9}$$

The unit of modulus of rigidity coincides with that of stress.

In Fig. 3-10, AC is a principal plane through A, and the tensile stress on this plane is numerically equal to s_s. From the geometry of the deformed block, we obtain the following relationship between modulus of rigidity and modulus of elasticity:

$$G = \frac{E}{2(1 + m)} \tag{3-10}$$

where m denotes Poisson's ratio.

Example 3-5.   A structural material has the values $E = 8,000,000$ psi and $G = 3,200,000$ psi. Find the value of Poisson's ratio.
Solution.   By Eq. (3-10),

$$m = \frac{E}{2G} - 1 = \frac{8}{6.4} - 1 = \mathbf{0.25}$$

Alternatively, the modulus of rigidity is sometimes called the *shearing modulus of elasticity* and denoted by E_s.

3-5   Simple Statically Indeterminate Members.   A structural member is said to be *statically indeterminate* if it cannot be analyzed merely by applying the equations of equilibrium. This condition occurs when the number of unknown quantities exceeds the number of independent equations of equilibrium that can be developed.

Therefore, to analyze this type of member, it is necessary to supplement the equations of equilibrium with a system of equations that are formed by applying all known characteristics of the deformation of the member.

Example 3-6. A Copperweld wire with an external diameter of $\frac{3}{8}$ in. has a steel core of $\frac{1}{4}$-in. diameter. What are the stresses in the steel and copper under an axial tensile load of 600 lb? The values of E are 30,000,000 and 15,000,000 psi for steel and copper, respectively. The two materials are connected in such manner that the wire functions as a unit.

Solution. Let the subscripts t and c refer to steel and copper, respectively. The areas are as follows:

$$A_{\text{gross}} = 0.7854(\tfrac{3}{8})^2 = 0.1104 \text{ sq in.}$$
$$A_t = 0.7854(\tfrac{1}{4})^2 = 0.0491 \text{ sq in.}$$

Then
$$A_c = 0.1104 - 0.0491 = 0.0613 \text{ sq in.}$$

The total load carried by the composite member can be expressed in this form:

$$A_t s_t + A_c s_c = 600$$

or
$$0.0491 s_t + 0.0613 s_c = 600 \qquad (a)$$

The known characteristic concerning the deformation is the following: The two elements (steel core and copper skin) elongate the same amount. Applying Eq. (3-6), we obtain

$$\Delta L = \frac{s_t L}{E_t} = \frac{s_c L}{E_c}$$

$$\therefore \; s_t = s_c \frac{E_t}{E_c} = 2s_c$$

Equation (a) can now be rewritten as

$$0.0491 \times 2s_c + 0.0613 s_c = 600 \qquad \text{or} \qquad 0.1595 s_c = 600$$

Then
$$s_c = \textbf{3760 psi} \qquad s_t = 2 \times 3760 = \textbf{7520 psi}$$

Example 3-7. The three bars in Fig. 3-11 are tied together securely to function as a composite compression member. The bars are all $\frac{3}{4}$ in. thick. A load Q is to be applied to the cap plate at such location that all three bars contract uniformly. On the basis of the following data, determine the allowable value of Q and the location of its line of action.

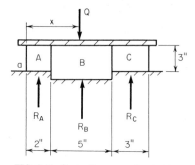

FIG. 3-11. Composite compression member.

Bar	Length, in.	E, psi	Allowable stress, psi
A	3	30,000,000	20,000
B	4	10,000,000	6,000
C	3	15,000,000	8,000

Solution. Let R_A, R_B, and R_C denote the resultant reactions on the respective bars. These are also the axial forces in the bars. It is necessary to identify the element that limits the capacity of the member. The known characteristic concerning the deformation is the following: All three elements contract the same amount. Equation (3-6) therefore yields the following:

$$\Delta L = \frac{s_A L_A}{E_A} = \frac{s_B L_B}{E_B} = \frac{s_C L_C}{E_C}$$

where s denotes the true stress in the element. Then

$$\frac{s_A}{s_B} = \frac{30}{10}\frac{4}{3} = 4 \qquad \text{and} \qquad \frac{s_C}{s_B} = \frac{15}{10}\frac{4}{3} = 2$$

The stresses therefore have the following proportions:

	True stress	Allowable stress
s_A	4	3.33
s_B	1	1.00
s_C	2	1.33

The ratio of relative allowable stress to relative true stress is lowest for C, and this element therefore limits the capacity of the composite member.

When Q attains its limiting value, the stresses are as follows:

$$s_C = 8000 \text{ psi}$$
$$s_A = 8000 \times \tfrac{4}{2} = 16,000 \text{ psi} \qquad s_B = 8000 \times \tfrac{1}{2} = 4000 \text{ psi}$$

The corresponding axial forces are as follows:

$$R_A = s_A A_A = 16,000(2 \times 0.75) = 24,000 \text{ lb}$$
$$R_B = 4000(5 \times 0.75) = 15,000 \text{ lb}$$
$$R_C = 8000(3 \times 0.75) = 18,000 \text{ lb}$$

Then
$$Q = R_A + R_B + R_C = \mathbf{57,000 \text{ lb}}$$

The resultant reactions are concentrated at the centers of the bars. Let x denote the distance from the left edge a to the line of action of Q, as shown in Fig. 3-11.

$$\Sigma M_a = Qx - (R_A \times 1 + R_B \times 4.5 + R_C \times 8.5) = 0$$

Substituting numerical values, we obtain

$$x = \mathbf{4.29 \text{ in.}}$$

Example 3-8.   The bar in Fig. 3-12a is rigid. It is pinned to a support at the left end and hung from rods 1 and 2 at the indicated locations. The

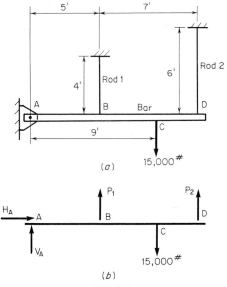

(a)

(b)

FIG. 3-12

cross-sectional areas of rods 1 and 2 are 1.0 and 1.4 sq in., respectively. The rods are both made of a material having a modulus of elasticity of 18,000,000 psi. Find the reactions and the vertical displacement of point C resulting from the 15,000-lb load.

Solution.   Let P_1 and P_2 denote the reactions of rods 1 and 2, respectively. The free-body diagram of the bar is shown in Fig. 3-12b. There are four reactions, and the equations of equilibrium yield the following:

$$\Sigma F_H = 0 \qquad\qquad \therefore\ H_A = 0 \qquad\qquad (a)$$
$$\Sigma F_V = 0 \qquad \therefore\ V_A + P_1 + P_2 = 15{,}000 \qquad (b)$$
$$\Sigma M_A = 15{,}000 \times 9 - (5P_1 + 12P_2) = 0$$
$$\therefore\ 5P_1 + 12P_2 = 135{,}000 \qquad\qquad (c)$$

When the load is applied at C, the bar pivots about A. Since the bar is considered to be infinitely rigid, its longitudinal axis remains straight, and therefore each hanger elongates an amount that is directly proportional to its distance from A. Then

$$\frac{\Delta L_2}{\Delta L_1} = \frac{12}{5}$$

Applying Eq. (3-6), we obtain

$$\frac{\Delta L_2}{\Delta L_1} = \frac{P_2}{P_1}\frac{L_2}{L_1}\frac{A_1}{A_2} = \frac{12}{5}$$

Then

$$\frac{P_2}{P_1} = \frac{12}{5}\frac{4}{6}\frac{1.4}{1.0} = 2.24 \qquad\qquad (d)$$

From Eq. (c), $5P_1 + 12 \times 2.24P_1 = 135{,}000$
Solving, $P_1 = 4235$ lb $P_2 = 2.24 \times 4235 = 9486$ lb
From Eq. (b), $V_A = 1279$ lb

Applying the data for rod 2, we have

$$\Delta L_2 = \frac{P_2 L_2}{A_2 E} = \frac{9486 \times 72}{1.4 \times 18{,}000{,}000} = 0.0271 \text{ in.}$$

By proportion,

$$\text{Displacement of } C = 0.0271 \times \tfrac{9}{12} = \mathbf{0.0203 \text{ in.}}$$

3-6   Thermal Stresses.   If a body is free to expand or contract, its volume fluctuates as the temperature varies. In our study, we shall be concerned solely with the change in the length of the member. Within the usual temperature range, the change in length may be considered to be directly proportional to the change in temperature. Let

L = original length, in.
ΔL = change in length, in.
ΔT = change in temperature, °F

Then $$\Delta L = cL \,\Delta T \tag{3-11}$$

where the constant of proportionality c is called the *coefficient of thermal expansion*. Its unit is in. per in.-°F, or simply per °F.

Example 3-9. A brass member supported by rods has a length of 20 ft at a temperature of 90°F. How much will it contract when the temperature drops to 40°F if the coefficient of thermal expansion is 0.0000092 per °F?

Solution
$$\Delta L = 0.0000092 \times 20 \times 12 \times 50 = \mathbf{0.11 \text{ in.}}$$

If the expansion or contraction of a member in response to temperature variations is restricted by contiguous bodies, these restraining forces must be taken into account in designing the member. The stresses caused by such restraint are known as *thermal stresses*. The restraining forces can be evaluated by modifying the sequence of events, in this manner: Allow the member to undergo its natural deformation, and then apply forces to reduce that deformation to its true value. Since ΔL is minuscule in relation to L, all calculations are based on the original length.

Let P denote the axial restraining force and ΔL the true change in length. By combining Eqs. (3-6) and (3-11), we obtain the following:

$$\Delta L = cL \,\Delta T - \frac{PL}{AE}$$

or $$\frac{\Delta L}{L} = c \,\Delta T - \frac{P}{AE} = c \,\Delta T - \frac{s}{E} \tag{3-12}$$

Example 3-10. A bar 4 ft long is set snugly between two walls when the temperature is 50°F. What is the stress in the bar when the temperature is

90°F (*a*) if the walls are rigid; (*b*) if the walls have yielded 0.008 in.? Use $E = 17 \times 10^6$ psi and $c = 9.3 \times 10^{-6}$ per °F.

Solution. PART *a*

$$\Delta T = 40°\text{F} \qquad \Delta L = 0$$

Substituting in Eq. (3-12),

$$0 = 9.3 \times 10^{-6} \times 40 - \frac{s}{17 \times 10^6}$$

$$s = \textbf{6320 psi}$$

PART *b*

$$\Delta T = 40°\text{F} \qquad \Delta L = 0.008 \text{ in.}$$

$$\frac{0.008}{48} = 9.3 \times 10^{-6} \times 40 - \frac{s}{17 \times 10^6}$$

$$s = \textbf{3490 psi}$$

3-7 Thin Circular Hoops. If a hollow circular member has a thickness that is extremely small in relation to its diameter, the member may be regarded as composed of infinitesimally small elements of negligible curvature. Consequently, Eq. (3-11) is applicable in determining the change in the circumferential length of this member caused by a change in temperature. Since the circumferential length is directly proportional to the diameter, Eq. (3-11) yields the following:

$$\Delta D = cD \, \Delta T \tag{3-13}$$

where D is the original mean diameter and ΔD is the increase in diameter. In practice, D may be taken as the internal diameter without significantly affecting the calculated value of ΔD.

Example 3-11.   A thin steel hoop having an internal diameter of 35.994 in. is to be heated to enable it to fit over a cylinder of precisely 36-in. diameter, with a clearance of 0.08 in. all around. If $c = 6.5 \times 10^{-6}$ per °F, by what amount must the temperature of the hoop be elevated?

Solution

$$\Delta D = 36.000 + 2 \times 0.080 - 35.994 = 0.166 \text{ in.}$$

By Eq. (3-13),

$$\Delta T = \frac{\Delta D}{cD} = \frac{0.166}{6.5 \times 10^{-6} \times 35.994} = \frac{0.166 \times 10^6}{6.5 \times 35.994} = \mathbf{710°F}$$

The hollow cylinder in Fig. 3-13a sustains a uniform internal pressure p that is directed radially outward. Since the cylinder yields slightly, its circumferential length increases, and this increase in length signifies the presence of tension acting in a transverse plane. It is therefore necessary to evaluate this tension. Let

  L = length of cylinder
  D = internal diameter
  t = thickness of shell

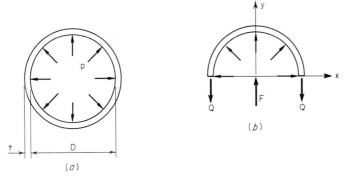

FIG. 3-13. Cylinder under radial pressure.

Consider the cylinder to be cut along a diametral plane, as shown in Fig. 3-13b, and establish the x and y axes indicated. To find the resultant F of the distributed force acting on the semicylindrical surface, take the force $p\,dA$ acting on an elemental area, resolve this force into its x and y components, and then integrate across the entire surface. The results are as follows:

$$F_x = 0 \qquad F_y = pDL$$

(More simply, the x and y components of $p\,dA$ equal p times the area dA as projected onto the y and x axes, respectively.)

As indicated in Fig. 3-13b, the force F is resisted by tensile forces Q in the cylinder shell. Set $Q = F_y/2$, giving

$$Q = \frac{pDL}{2} \qquad (3\text{-}14)$$

Let s denote the stress corresponding to Q. For a thin-walled cylinder (one for which $t/D \leqslant \frac{1}{15}$), this stress may be considered to be uniform across the thickness of the shell. Divide Q by its area tL, giving

$$s = \frac{pD}{2t} \qquad (3\text{-}15)$$

The force Q and its stress s are referred to as *hoop tension* and *hoop stress*, respectively.

Since the thin-walled cylinder may be regarded as composed of infinitesimally small elements of negligible curvature, Eq. (3-6) is applicable, and it yields the following:

$$\Delta D = \frac{sD}{E} \qquad (3\text{-}16a)$$

where s denotes the hoop stress. Replace s with the expression in Eq. (3-15), giving

$$\Delta D = \frac{pD^2}{2tE} \qquad (3\text{-}16b)$$

If the radial pressure p is directed inward, the hoop stress is compressive and the diameter decreases.

Example 3-12. A cylinder having an internal diameter of 5 ft and a thickness of $\frac{3}{4}$ in. is subjected to a radial outward pressure of 180 psi. Find the hoop stress and the increase in diameter, using $E = 30 \times 10^6$ psi.
Solution. By Eq. (3-15),

$$s = \frac{180 \times 60}{2 \times \frac{3}{4}} = \textbf{7200 psi}$$

By Eq. (3-16a),

$$\Delta D = \frac{7200 \times 60}{30 \times 10^6} = \textbf{0.0144 in.}$$

3-8 Elastic Design and Strength Design. As stated in Art. 3-4, the elastic limit of a structural material is the limiting stress that may be induced in a body without imparting a permanent deformation (or *set*) to the body. The proportional limit of a material is the limiting stress at which strain remains directly proportional to stress.

When a bar of structural steel is subjected to a gradually increasing tensile stress, a point is reached at which the member undergoes a vast increase in deformation without any increase in stress. This behavior is referred to as *yielding*, and the stress at which yielding occurs is termed the *yield-point stress*. For structural steel, the elastic limit, proportional limit, and yield-point stress are all close in value, and in practice they may be considered to be coincident. The deformation of a steel bar in compression is similar to that in tension.

Consider that a structure is subjected to a gradually increasing load until it collapses. When the yield-point stress is attained, the structure is said to be in a state of *initial yielding*. The load that exists when failure impends is termed the *ultimate load*. In *elastic design*, a structure is considered to be loaded to capacity when it attains initial yielding, on the theory that deformation beyond this point would annul the utility of the structure. In *strength* (or *ultimate-strength*) *design*, on the other hand, it is recognized that a structure may be loaded beyond initial yielding if those parts of the structure where the stress is below the yield-point stress are capable of resisting the tendency of the structure to yield and of supporting this incremental load. Collapse occurs when these requirements are no longer satisfied.

Thus, elastic design is concerned with an *allowable stress*, which equals the yield-point stress divided by an appropriate *factor of safety*. In contrast, strength design is concerned with an *allowable load*, which equals the ultimate load divided by an appropriate factor called the *load factor*.

The load to be supported by a structure is termed the *service* or *working* load. Strength design is referred to as *plastic design* when applied to steel. An alternative expression for elastic design is *working-stress design*.

In analyzing reinforced-concrete beams, we shall apply strength design, as the present design code requires. However, in analyzing steel beams, we shall confine ourselves to elastic design.

Many design codes that are ostensibly based on elastic design implicitly recognize the behavior of a body stressed beyond initial yielding. Consequently, there is no sharp line of demarcation between the two methods of design.

PROBLEMS

In solving the following problems, apply the values recorded in the accompanying table where they are relevant.

	E, psi	c, per $°F$
Steel	30×10^6	6.5×10^{-6}
Copper	15×10^6	9.0×10^{-6}
Aluminum	10×10^6	13.0×10^{-6}

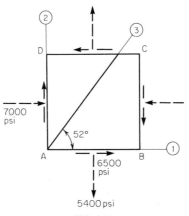

FIG. 3-14

3-1. With reference to the block in Fig. 3-14, determine the following: the principal stresses and maximum shearing stress at A, the angle between the plane of f_{min} and plane 1, the stresses on plane 3, and the absolute value of the shearing stress on a plane of pure shear, denoted as plane 4.

ANS. $f_{max} = 8183$ psi tension; $f_{min} = 9783$ psi compression; $s_{s,max} = 8983$ psi; $\alpha = -66°49'$; $f_3 = 4007$ psi tension; $s_{s3} = 7588$ psi; $s_{s4} = 8947$ psi

3-2. The block in Fig. 3-15 sustains the indicated stresses. If $E = 12 \times 10^6$ psi and the strain ϵ_x in the x direction is found to be 49.2×10^{-6}, what is the value of Poisson's ratio?   ANS. 0.290

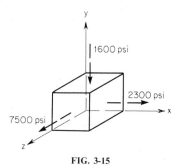

FIG. 3-15

3-3. A steel hanger of 7-ft length is to carry an axial load of 40 kips. If the tensile stress is restricted to 15,000 psi and the elongation caused by this load is restricted to 0.04 in., what is the minimum cross-sectional area the member can have?

ANS. 2.80 sq. in.

3-4. The homogeneous bar in Fig. 3-16 is in the form of a right circular cone. Let L denote the length, R the radius at the base of cone, w the specific weight of the material, and E the modulus of elasticity. If the bar is suspended vertically in the manner shown, prove that the increase in length of the part AB resulting from the weight of the member is

$$\Delta L_{AB} = \frac{wh^2}{6E}$$

Note that this expression is independent of R.

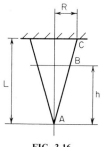

FIG. 3-16

3-5. The homogeneous bar in Fig. 3-17 has the following cross-sectional areas: from A to C, 2.0 sq. in.; from C to D, 1.5 sq in. The bar is connected to two rigid supports, and a total load of 18 kips is applied concentrically to the bar at section B. Find the reactions at A and D.   ANS. $R_A = 12$ kips; $R_D = 6$ kips

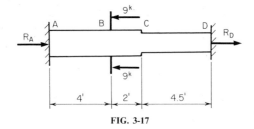

FIG. 3-17

HINT: The following system of equations applies to the member:

$$\Sigma F_H = R_A + R_D - 18 = 0$$
$$\Delta L_{AB} + \Delta L_{BC} + \Delta L_{CD} = 0$$

Consider an elongation positive and a contraction negative. What is the internal force in AB and BD?

3-6. A $\frac{1}{2}$-in. Copperweld wire has a steel core of $\frac{5}{16}$-in. diameter. What tensile load applied to the member will produce a stress of 3800 psi in the copper?

ANS. 1038 lb

3-7. The rigid bar in Fig. 3-18 carries a vertical load of 20,000 lb at the indicated location and is supported by three steel rods in the manner shown. The rods have the following relative values of cross-sectional areas: $A_1 = 1.25$, $A_2 = 1.20$, $A_3 = 1.00$, where the subscripts correspond to the rod numbers. Find the tensile force in each rod and locate the center of rotation of the bar.

ANS. $P_1 = 11,814$ lb; $P_2 = 5097$ lb; $P_3 = 3089$ lb; center of rotation lies 31.4 ft to right of rod 1.

HINT: Draw the frame in its deformed position and establish the relationship

$$\frac{\Delta L_1 - \Delta L_2}{\Delta L_2 - \Delta L_3} = \frac{6}{10}$$

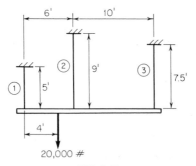

20,000 #

FIG. 3-18

3-8. A heavy steel strut 12 ft long was set snugly between two walls. It was later found that the strut had expanded 0.018 in. and was stressed to 3400 psi in compression. What was the temperature rise during this interval?   ANS. 36.6°F

3-9. An aluminum and a steel bar have the following data: length, 15 in. for aluminum and 9 in. for steel; cross-sectional area, 1.2 for aluminum and 1.5 for steel. The bars were set end to end between two walls exactly 2 ft apart, with adequate restraint against buckling. If the temperature rose 80°F and the walls yielded 0.01 in. altogether, what was the compressive stress in each bar?

ANS. Aluminum, 5910 psi; steel, 4730 psi

3-10. The steel and copper bars in Fig. 3-19 have the same length and they are connected to rigid plates at top and bottom. The cross-sectional areas are 1.2 sq in. for the steel and 1.5 sq in. for the copper. If the temperature increases 40°F after the built-up member is formed, what are the thermal stresses in the bars?

ANS. Steel, 1150 psi tension; copper, 920 psi compression

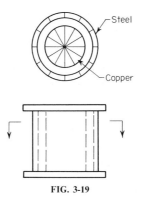

FIG. 3-19

3-11. With reference to Fig. 3-19, demonstrate that the coefficient of thermal expansion of the composite member consisting of the steel and copper bars is

$$c = \frac{c_s A_s E_s + c_c A_c E_c}{A_s E_s + A_c E_c}$$

where the subscripts s and c refer to steel and copper, respectively. Thus, the coefficient of the composite member is a weighted average, obtained by weighting the coefficients of the constituents by their respective AE values.

3-12. A circular steel ring 2 in. thick was heated and allowed to shrink over a rigid cylinder having a diameter of 54.00 in. When the ring had cooled, the radial pressure between ring and cylinder was 1280 psi. What was the original internal diameter of the ring?   ANS. 53.969 in.

4

PRESSURE VESSELS—
RIVETED AND WELDED
CONNECTIONS

4-1 Stresses in Cylindrical Vessels. Consider that a thin-walled cylindrical vessel contains a fluid at rest, and assume that across a length L the fluid exerts a uniform outward radial pressure upon the cylinder. As we found in Art. 3-7, the shell of the vessel sustains tension that acts in a transverse plane. In accordance with Eqs. (3-14) and (3-15),

$$Q = \frac{pDL}{2} \quad \text{and} \quad s = \frac{pD}{2t}$$

where Q = hoop tension
    s = hoop stress
    p = fluid pressure
    D = diameter of vessel
    t = thickness of shell

Example 4-1. A pipe 4 ft in diameter is subjected to a fluid pressure of 175 psi. If the pipe is $\frac{3}{8}$ in. thick, what is the hoop stress in the pipe?

Solution

$$s = \frac{175 \times 48}{2 \times \frac{3}{8}} = \textbf{11,200 psi}$$

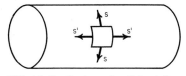

FIG. 4-1. Tensile stresses in cylinder shell.

Figure 4-1 shows a horizontal cylindrical vessel with closed ends. Assume that the pressure of the confined fluid is uniform throughout the vessel, and let p denote this pressure. Since the fluid exerts an outward thrust on the closed ends of the vessel, the shell also sustains a tensile stress s' in the longitudinal direction. Dividing the thrust on a closed end by the transverse area of the vessel, we obtain

$$s' = \frac{p\pi D^2/4}{\pi Dt} = \frac{pD}{4t} \tag{4-1}$$

In addition to the tensile stresses s and s', the shell is subjected to a radial compressive stress that ranges from p at the interior surface to 0 at the exterior surface.

If the cylinder is built up of individual parts, the bands that serve as the ligature of the member are also subjected to hoop tension. This condition is illustrated by a wood-stave pipe, where steel hoops are placed at regular intervals to clamp the staves together. If we equate L in Eq. (3-14) with the distance between successive hoops, we obtain the tension in the hoop.

Example 4-2.   The water in a 6-ft-diameter wood-stave penstock is under a pressure of 50 psi. If the hoops are spaced 6 in. on centers and the allowable tensile stress is 16,000 psi in the gross section, what must be the size of the hoops?

Solution

$$Q = \tfrac{1}{2} \times 50 \times 72 \times 6 = 10{,}800 \text{ lb}$$
$$\text{Area required} = 10{,}800/16{,}000 = 0.675 \text{ sq in.}$$
$$\therefore \text{ Use } \textbf{1-in.-diameter} \text{ hoops.} \qquad \text{Area} = 0.785 \text{ sq in.}$$

Example 4-3.   A wood-stave penstock of 4-ft diameter is wrapped with $\tfrac{7}{8}$-in. steel hoops at a spacing of $4\tfrac{1}{2}$ in. The allowable stress in tension is 12,000 psi at the root of the threaded ends of the hoops, and the net area of a $\tfrac{7}{8}$-in. hoop is 0.419 sq in. What fluid pressure can be applied on the basis of tension in the hoops?

Solution

$$p = \frac{2Q}{DL} = \frac{2 \times 0.419 \times 12,000}{48 \times 4.5} = \textbf{46.6 psi}$$

4-2   Deformation of Cylindrical Vessels.   In analyzing the deformation of a closed cylindrical vessel, the effects of the radial compressive stress p may be disregarded. By referring to Fig. 4-1 and taking into account the lateral strain caused by each stress, it is seen that stress s tends to increase the diameter and decrease the length, and stress s' tends to increase the length and decrease the diameter.

Let D, L, and V denote the original diameter, length, and volume of the vessel, respectively. Let m denote Poisson's ratio, which is defined in Art. 3-4. The deformations of the vessel are as follows:

$$\Delta D = \frac{D}{E}(s - ms') = \frac{pD^2}{4tE}(2 - m) \qquad (4\text{-}2a)$$

$$\Delta L = \frac{L}{E}(s' - ms) = \frac{pDL}{4tE}(1 - 2m) \qquad (4\text{-}2b)$$

If the deformation of the closed ends is disregarded, the increase in volume is

$$\Delta V = \frac{pDV}{tE}(1.25 - m) \qquad (4\text{-}2c)$$

Example 4-4.   A cylindrical tank is 63 in. in diameter, 12 ft long, and 0.50 in. thick. The tank is subjected to an internal fluid pressure of 280 psi. What is the increase in capacity of the tank resulting from its deformation? Use $m = 0.25$ and $E = 30 \times 10^6$ psi, and express the answer in cu in.

Solution

$$V = \frac{\pi}{4}D^2L = 0.785 \times 63^2 \times 144 = 449,000 \text{ cu in.}$$

$$\Delta V = \frac{280 \times 63 \times 449,000}{0.50 \times 30,000,000}(1.25 - 0.25) = \textbf{528 cu in.}$$

4-3   Prestressed Concrete Pipes.   Concrete is a desirable structural material because of the ease with which it can be fashioned, but its low tensile strength constitutes a serious obstacle. In the construction of pressure vessels, this obstacle is surmounted by inducing compressive stresses in the concrete before the fluid is admitted to the vessel, and the concrete is then said to be *prestressed*. The procedure consists of enveloping the concrete shell with high-strength steel wires and applying tensile forces to the wires, thereby compressing the enclosed concrete. The member is then encased in mortar to insulate the steel. When the fluid is admitted, the concrete shell and steel wires act in combination to resist the outward thrust of the fluid. Thus, the steel wires reduce the tensile stress in the concrete resulting from fluid pressure, and the prestress in the concrete serves to regulate the final stress in that material.

We shall confine our study to thin-walled cylindrical pressure vessels. Let

A_s = cross-sectional area of steel wire

D = internal diameter of vessel

t = thickness of shell

h = center-to-center spacing of wires (or pitch of wire, if wrapped spirally)

n = ratio of modulus of elasticity of steel to that of concrete
   $= E_s/E_c$

s_c = prestress in concrete

s_c' = hoop stress in concrete due to fluid pressure

s_c'' = final stress in concrete

To represent the corresponding stresses in the steel, we shall replace the subscript c with s. Tensile stresses will be considered positive, and compressive stresses negative. For simplicity, we shall disregard the modification of stress engendered by shrinkage of the concrete under sustained load. The basic equations are as follows:

$$s_c + s_c' = s_c''$$
$$s_s + s_s' = s_s'' \tag{4-3}$$

Consider a portion of the pipe h in. long. Figure 4-2 shows the resultant internal forces on this body caused by prestressing the member.

$$T = -C$$

But $\qquad T = A_s s_s \qquad$ and $\qquad C = hts_c$

$$\therefore s_c = -\frac{T}{ht} = -\frac{A_s s_s}{ht} \tag{4-4}$$

These prestresses are shown in the stress diagram, Fig. 4-3.

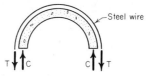

FIG. 4-2. Prestress condition.

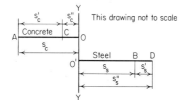

FIG. 4-3. Stresses in concrete shell and steel wire.

The outward thrust of the confined fluid is resisted jointly by the concrete and steel.

$$\therefore \; pDh = 2(A_s s_s' + ht s_c')$$

Since the steel ring and concrete shell have approximately the same diameter, we may consider their strains caused by fluid pressure to be equal.

$$\epsilon = \frac{s_s'}{E_s} = \frac{s_c'}{E_c} \qquad \therefore \; s_s' = n s_c'$$

Substituting in the preceding equation, we obtain

$$s_c' = \frac{pDh}{2(nA_s + ht)}$$

$$s_s' = \frac{pDh}{2(A_s + ht/n)}$$

(4-5)

The equation for s_c' can readily be memorized if we conceive the steel ring to be supplanted with one of concrete having an area nA_s. The denominator would then represent the total area of concrete resisting the fluid thrust.

Since the pipe is in equilibrium,

$$A_s s_s'' + ht s_c'' = \tfrac{1}{2} pDh$$

(4-6)

Example 4-5. A pipe 6 ft in diameter is cast as a concrete shell 4 in. thick and is then wrapped with wires $\frac{1}{4}$ in. in diameter under a tensile stress of 40,000 psi. The wires are spaced $1\frac{1}{2}$ in. on centers. Determine the stresses in the concrete and steel when the fluid pressure is 35 psi. Use $n = 12$.

Solution

$$s_c = -\frac{0.0491 \times 40,000}{1.5 \times 4} = -327 \text{ psi}$$

$$s'_c = \frac{35 \times 72 \times 1.5}{2(12 \times 0.0491 + 1.5 \times 4)} = 287 \text{ psi}$$

$$s'_s = 12 \times 287 = 3440 \text{ psi}$$
$$s''_c = -327 + 287 = \mathbf{-40 \text{ psi}}$$
$$s''_s = 40,000 + 3440 = \mathbf{43,440 \text{ psi}}$$

These results can be verified by substituting in Eq. (4-6).

Example 4-6. A water pipe of 5-ft inside diameter will operate under a pressure of 50 psi. The pipe is cast of concrete 3 in. thick and then wound spirally with $\frac{1}{4}$-in. wire under tension. If the steel stress is not to exceed 30,000 psi and the concrete stress in compression is to be 100 psi minimum, what should be the pitch of the wire? What initial tension is required? Use $n = 10$.

Solution. Greater clarity is achieved if we arrange the variables schematically in the manner indicated and insert the value of each variable as it becomes known.

s_c	s'_c	$s''_c = -100$ psi
s_s	s'_s	$s''_s = 30,000$ psi
$p = 50$ psi	$h = ?$	$T = ?$

By rearranging Eq. (4-6), we obtain

$$h = \frac{2A_s s''_s}{pD - 2ts''_c} = \frac{2 \times 0.0491 \times 30,000}{50 \times 60 + 2 \times 3 \times 100} = \mathbf{0.82 \text{ in.}}$$

From Eq. (4-5),

$$s'_s = \frac{50 \times 60 \times 0.82}{2(0.0491 + 0.82 \times \frac{3}{10})} = 4170 \text{ psi}$$

$$s_s = 30,000 - 4170 = 25,830 \text{ psi}$$
$$T = 0.0491 \times 25,830 = \mathbf{1270 \text{ lb}}$$

Check

$$s_c' = \frac{4170}{10} = 417 \text{ psi}$$
$$s_c = -100 - 417 = -517 \text{ psi}$$
$$C = -0.82 \times 3 \times 517 = -1270 \text{ lb} = -T$$

4-4   Stresses in Riveted Joints.   Consider that a plate A is subjected to a concentric force P and that it must transmit this force to another plate B. In Fig. 4-4a, this force is transmitted directly by means of the four rivets shown, and this type of connection is known as a *lap* joint. In Fig. 4-4e, the force is transmitted through the media of the auxiliary plates C and D. This type of connection is known as a *butt* joint, and the auxiliary plates are referred to as *cover, strap,* or *splice* plates. A riveted joint behaves as a chain, and each link must be capable of transmitting the applied force. The rivets are considered to be equally stressed if the force transmitted through the joint is concentric with respect to the rivet group. Friction between the connected members is usually ignored.

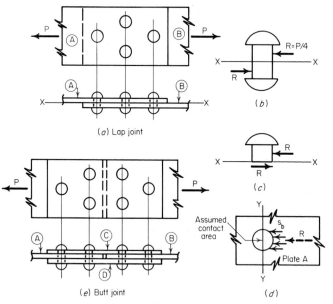

(a) Lap joint

(b)

(c)

(e) Butt joint

(d)

FIG. 4-4

Assume that the rivets remain stationary as load is applied. Plate *A* in Fig. 4-4*a* moves slightly to the left, engaging each rivet along its right half, and plate *B* moves in the opposite direction. The resultant forces acting on the upper and lower parts of a rivet are shown in Fig. 4-4*b*. If the rivet is severed along the contact surface of the plates, as shown in Fig. 4-4*c*, it is seen that there is an internal shearing force *R* on the transverse section at this plane. This force is assumed to be uniformly distributed across the area. Let *d* denote the rivet diameter, A_s the cross-sectional area, and s_s the shearing stress. Then

$$s_s = \frac{R}{A_s} = \frac{4R}{\pi d^2} \qquad (4\text{-}7)$$

When the opposing forces tend to shear a rivet along one plane, as in Fig. 4-4*a*, the rivet is said to be in *single shear*; when they tend to shear a rivet along two planes, as in Fig. 4-4*e*, the rivet is said to be in *double shear*. A rivet in double shear has a shearing capacity twice as great as that of a rivet in single shear.

The stress that the plate exerts on the rivet along their surface of contact is termed the *bearing* stress; it is denoted by s_b. Figure 4-4*d* is a view parallel to the longitudinal axis of the rivet shown in Fig. 4-4*b*, and the diametral plane *YY* through the rivet is normal to the action line of *R*. To simplify the calculation of bearing stress, the semicylindrical surface of contact of the plate and rivet is replaced with its projection on *YY*, and the bearing stress is assumed to be uniform across this projected surface. Therefore, if *t* denotes the thickness of plate,

$$s_b = \frac{R}{tD} \qquad (4\text{-}8)$$

In addition to the shearing and bearing stress, a rivet also sustains bending stress, but this is negligible in the case of a short rivet.

The presence of rivet holes impairs the tensile strength of the connected members. To investigate this matter, refer to Fig. 4-5*a*, which is a free-body diagram of the right end of plate *A* in Fig. 4-4*a*. Consider the member to be divided into the parts shown in Fig. 4-5*b* and *c*. Let *d'* denote the effective diameter of rivet hole, *w* the width of plate, and s_t and s_t' the tensile stress in the plate at *MM* and *NN*, respectively. The tensile stress is assumed to be uniform at each

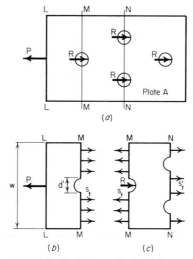

FIG. 4-5. Tensile stresses in riveted plate.

section, and therefore

$$s_t = \frac{P}{t(w - d')} \qquad s_t' = \frac{P - R}{t(w - 2d')}$$

It is necessary to compute the tensile stress in the plate at each transverse row of rivets, since the location of the critical section is not self-evident. For maximum efficiency, the rivets of a joint are usually arranged in a diamond formation, thereby approximately equating the rate at which the net plate area diminishes with the rate at which the tensile force is transferred. Rivet holes are generally made $\frac{1}{16}$ in. larger than the nominal rivet diameter, but the effective diameter of the hole is considered to be $\frac{1}{8}$ in. more than the rivet diameter as a means of allowing for damage to the adjacent metal. If the force transmitted through the joint is compressive and the rivets completely fill their holes, the gross area of the member may be applied in computing the compressive stress.

Ideally, a riveted joint should be so designed that all stresses simultaneously attain their allowable values as the load is gradually increased. The maximum force that the joint can transmit is termed its *capacity*, and the ratio of the capacity of the joint to the capacity of the gross section of the main member is referred to as the *efficiency* of the joint.

Figure 4-6, which shows a typical connection in a steel truss, illustrates certain definitions pertaining to riveted joints. Each truss member consists of two angle sections, with a space between the backs of angles to allow insertion of the gusset plate. Since the rivets in the horizontal member are not aligned along a transverse axis of that member, they are said to be *staggered*.

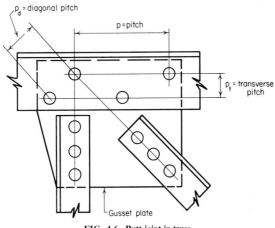

FIG. 4-6. Butt joint in truss.

In the following material, it is to be understood that all main members and connection plates are made of ASTM A36 steel, which has a yield-point stress of 36,000 psi, and that the connections are made with A502, Grade 1, hot-driven rivets. The Specification of the American Institute of Steel Construction provides the following allowable stresses: tensile stress in a main member or connection plate, 22,000 psi; bearing stress on a main member or connection plate, 48,600 psi; shearing stress in a rivet, 15,000 psi. The ratio of the center-to-center distance between any two adjacent rivets to the rivet diameter should not be less than 2.67, and preferably not less than 3. The minimum distance from center of rivet to the edge of any member is recorded in the accompanying table.

Rivet diameter, in.	Minimum edge distance, in.	
	At sheared edge	At rolled edge
$\frac{5}{8}$	$1\frac{1}{8}$	$\frac{7}{8}$
$\frac{3}{4}$	$1\frac{1}{4}$	1
$\frac{7}{8}$	$1\frac{1}{2}*$	$1\frac{1}{8}$
1	$1\frac{3}{4}*$	$1\frac{1}{4}$

* May be $1\frac{1}{4}$ in. at ends of beam connection angles.

Example 4-7.   The hanger in Fig. 4-7 is a 10- by $\frac{3}{8}$-in. steel plate, and it is connected to its support with nine $\frac{3}{4}$-in. rivets in single shear, as shown. The supporting member has the same section. What is the maximum vertical load that can be suspended? What is the efficiency of the joint?

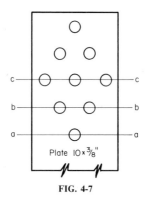

FIG. 4-7

Solution.   The effective diameter of rivet holes is $\frac{7}{8}$ in. Let P denote the allowable load on the member, with the subscripts t, b, and s denoting the criteria of tension, bearing, and shear, respectively. The suspended load displaces the hanger downward, and each rivet therefore exerts its reaction directly *above* the horizontal diametral plane through the rivet. Therefore, the full load acts at section aa, eight-ninths of the full load acts at bb, and two-thirds of the full load acts at cc. Applying the allowable stresses recorded above, we obtain the following:

Rivet capacity in shear $= 0.4418 \times 15{,}000 = 6630$ lb

$$\therefore\ P_s = 9 \times 6630 = 59{,}700 \text{ lb}$$
$$P_b = 9 \times \tfrac{3}{8} \times \tfrac{3}{4} \times 48{,}600 = 123{,}000 \text{ lb}$$

At aa: $P_t = (10 - 0.88) \times \tfrac{3}{8} \times 22{,}000 = 75{,}200$ lb

At bb: $\tfrac{8}{9}P_t = (10 - 1.75) \times \tfrac{3}{8} \times 22{,}000 = 68{,}100$ lb
$$P_t = 76{,}600 \text{ lb}$$

At cc: $\tfrac{2}{3}P_t = (10 - 2.63) \times \tfrac{3}{8} \times 22{,}000 = 60{,}800$ lb
$$P_t = 91{,}200 \text{ lb}$$

Selecting the lowest value, we find that

$$\text{Allowable load} = \textbf{59,700 lb}$$
$$\text{Capacity of gross plate area} = 10 \times \tfrac{3}{8} \times 22{,}000 = 82{,}500 \text{ lb}$$
$$\text{Joint efficiency} = 59{,}700/82{,}500 = \textbf{72 percent}$$

Example 4-8. Two 13- by $\tfrac{5}{8}$-in. steel plates are to be connected end to end by riveting them between two splice plates of 13-in. width. The plates are in tension. Design and detail a splice of maximum efficiency, using $\tfrac{7}{8}$-in. rivets.

Solution. Refer to Fig. 4-8, and consider the forces that the rivet at section aa exerts on the plates. The force on the main plate acts to the right of aa,

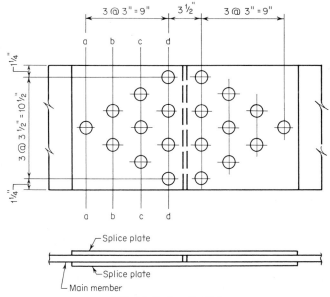

FIG. 4-8. Design of butt joint.

and the forces on the splice plates act to the left of that section. Therefore, the main plate carries its full load at *aa*. We shall design a splice for the maximum allowable load, which is the tensile capacity of a section 13 in. wide containing one rivet hole having an effective diameter of 1 in.

$$\text{Design load} = (13 - 1) \times \tfrac{5}{8} \times 22{,}000 = 165{,}000 \text{ lb}$$
$$\text{Rivet capacity in shear} = 2 \times 0.6013 \times 15{,}000 = 18{,}040 \text{ lb}$$
$$\text{Rivet capacity in bearing} = \tfrac{5}{8} \times \tfrac{7}{8} \times 48{,}600 = 26{,}580 \text{ lb}$$
$$\text{Number of rivets required} = 165{,}000/18{,}040 = 9.1$$
$$\therefore \text{ Use 10 rivets. Force on each rivet} = 16{,}500 \text{ lb}$$

The minimum rivet pitch is $3 \times \tfrac{7}{8} = 2\tfrac{5}{8}$ in., but 3 in. is the standard spacing. The minimum edge distance at a rolled edge is $1\tfrac{1}{8}$ in. We shall try the rivet pattern shown in Fig. 4-8. The calculations for tensile stress in the plate are recorded in the accompanying table.

Section	Force in main plate, lb	÷	Area, sq in.	=	Stress, psi
aa	165,000		$12 \times \tfrac{5}{8} = 7.5$		22,000
bb	148,500		$11 \times \tfrac{5}{8} = 6.88$		21,600
cc	115,500		$10 \times \tfrac{5}{8} = 6.25$		18,500
dd	66,000		$9 \times \tfrac{5}{8} = 5.63$		11,700

The trial pattern is therefore satisfactory, and it now remains to find the thickness of the splice plates. On the left side of the joint, each splice plate receives its rivet force to the *left* of the rivet center line. Therefore, the splice plates carry their full loads at section *dd*, where their area is minimum. Thus, the maximum tensile stress in the splice plates occurs at this section. Let t denote the required thickness of each plate.

$$t = \frac{\tfrac{1}{2} \times 165{,}000}{(13 - 4) \times 22{,}000} = 0.42 \text{ in.}$$

$\therefore$ Make splice plates 13 by $\tfrac{7}{16}$ in. As we shall see in Art. 9-1, the rivets may be grouped more compactly than indicated.

4-5   Design of Welded Joints.   In *fusion welding*, two metal sections are united by depositing weld metal along a groove at a temperature sufficiently high to melt both the weld metal and the adjacent base metal. The ensuing cooling of the metal causes the members to coalesce.

Two members that have coplanar faces and join end to end are connected to one another by means of a *butt* weld. The profiles of standard forms of butt welds are presented in the *Steel Construction Manual* published by the American Institute of Steel Construction (AISC).

Other members are usually connected by means of a *fillet* weld, which is illustrated in Fig. 4-9a. The weld profile is considered to be an isosceles triangle. As shown in Fig. 4-9b and c, the length of the two equal sides of this triangle is taken as the weld size, and the altitude from the root of weld to the third side is termed the *throat*. The throat area is the product of the throat thickness and the effective length of weld.

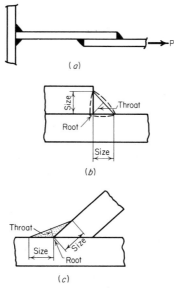

FIG. 4-9. Fillet welds.

Regardless of the manner in which a force is transmitted through a welded joint, it is assumed that the weld metal tends to fail in shear along its throat. If P denotes the transmitted force, A_t the throat area, and s the shearing stress on this area, the value of s is taken as

$$s = \frac{P}{A_t}$$

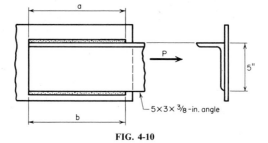

FIG. 4-10

Example 4-9. The 5-in. leg of the angle in Fig. 4-10 is to be connected to a $\frac{1}{2}$-in. gusset plate. The members are of A36 steel, and the angle will not sustain repeated variation of stress. Design a welded lap splice.

Solution. The connection will be designed for the tensile capacity of the angle, using an allowable axial stress of 22,000 psi. The properties of rolled steel sections are recorded in the AISC Manual, and the area of this angle is 2.86 sq in. Then

$$P_{max} = 2.86 \times 22,000 = 62,900 \text{ lb}$$

The required length of weld is found by computing the capacity of the connection in lb per linear in. (pli). In accordance with the AISC Specification, the minimum size of fillet weld that can be used with a $\frac{1}{2}$-in. plate is $\frac{3}{16}$ in., and the maximum size is $\frac{1}{16}$ in. less than the thickness of the angle. We shall therefore use a $\frac{5}{16}$-in. weld. The allowable shearing stress is 21,000 psi in the weld and 14,500 psi in the gusset plate. For a 1-in. length of weld,

Capacity of weld $= A_t s = 0.707 \times \frac{5}{16} \times 1 \times 21,000 = 4640$ pli
Capacity of plate $= \frac{1}{2} \times 1 \times 14,500 = 7250$ pli

Let L denote the required length of weld.

$$L = 62,900/4640 = 13.6 \text{ in.}$$

Since the member will not be subjected to repeated variation of stress, the AISC Specification allows us to disregard the eccentricity of the connection. We may therefore set $a = b = 7$ in.

PROBLEMS

4-1. A pipe is 5 ft in diameter and $\frac{1}{2}$ in. thick. What is the hoop stress in the pipe when the internal fluid pressure is 210 psi?   ANS. 12,600 psi

4-2. A wood-stave penstock of 6-ft diameter is wrapped with steel hoops and subjected to a fluid pressure of 55 psi. If the allowable tension in the hoops is 9600 lb, what is the maximum hoop spacing?   ANS. 4.85 in.

4-3. A cylindrical tank is 8 ft in diameter, 15 ft long, and $\frac{5}{8}$ in. thick. The tank is subjected to an internal fluid pressure of 200 psi. Find the increase in the diameter and length caused by this pressure, using $m = 0.25$ and $E = 30 \times 10^6$ psi.

ANS. 0.0430 in.; 0.0230 in.

4-4. A 6-ft pipe having a concrete shell 3 in. thick is wrapped spirally with $\frac{1}{4}$-in. wire. The steel stress is restricted to 48,000 psi, and the compressive stress in the concrete is restricted to a range of 40 to 500 psi. If $n = 10$, what is the maximum fluid pressure that may be applied?   ANS. 43 psi

HINT: Applying the given stresses, find s_s. Then calculate the pitch h of the wire.

4-5. A pipe of 64-in. diameter has a concrete shell of 3.3 in. and is wrapped spirally with $\frac{1}{4}$-in. wire under a tension of 1800 lb. If the fluid pressure is 35 psi, what pitch of wire is required to prevent tension in the concrete? Use $n = 9$.   ANS. 1.69 in.

5

STRESSES IN BEAMS

5-1 Definitions. A *beam* is a member that is subjected to coplanar forces having lines of action that are normal to the longitudinal axis of the member. The plane of the forces is a plane of symmetry of the transverse section, as illustrated in Fig. 5-1. In practice, however, many members that are unsymmetrical or sustain forces that differ from those described are also referred to as beams. For convenience of terminology, it is usually assumed that the longitudinal axis of the beam is horizontal.

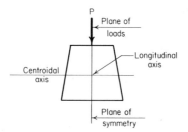

FIG. 5-1. Transverse section of beam.

Beams are classified according to the conditions of support. The member in Fig. 5-2 has two simple supports, and it is therefore termed a *simple*, or *simply supported*, beam. The distance between the supports is called the *span*. The member in Fig. 5-3 has two fixed supports, and it is therefore termed a *fixed*, or *fixed-ended*, beam. A beam that is fixed at one end but free or floating at the other end, as illustrated in Fig. 5-4, is referred to as a *cantilever* beam. The beam in Fig. 5-5, which overhangs its two supports and carries loads on the overhangs, is called an *overhanging* beam. If a beam has more than one span, as shown in Fig. 5-6, it is termed a *continuous* beam. These categories are not necessarily mutually exclusive, and combinations of these five basic beam types are possible, as illustrated by the member in Fig. 5-7.

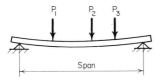

FIG. 5-2. Simple beam.

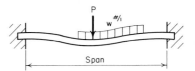

FIG. 5-3. Fixed beam.

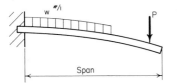

FIG. 5-4. Cantilever beam.

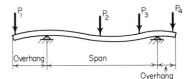

FIG. 5-5. Overhanging beam.

FIG. 5-6. Continuous beam.   FIG. 5-7

5-2 Reactions for Statically Determinate Beams. The first problem that arises in the design or investigation of a beam is to compute the reactions induced at the supports.

Example 5-1. Calculate the reactions for the beams shown in Figs. 5-8 to 5-12.

Solution. Since all loads are vertical, all reactions are either vertical forces or couples. As previously stated, we are concerned solely with the moment associated with a couple. By taking moments of all forces on the beam with respect to a support, we obtain an equation in which the number of unknowns is one less than the number of reactions. Subscripts will be used to identify the moment center.

PART *a*. (*Refer to Fig. 5-8.*)

$$\Sigma M_R = 20R_L - 3 \times 10 - 4 \times 6 = 0 \qquad \therefore R_L = \textbf{2.7 kips}$$
$$\Sigma M_L = 3 \times 10 + 4 \times 14 - 20R_R = 0 \qquad \therefore R_R = \textbf{4.3 kips}$$
$$\Sigma F = 2.7 + 4.3 - 3 - 4 = 0 \qquad \text{OK}$$

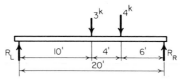

FIG. 5-8

PART *b*. (*Refer to Fig. 5-9.*) In accordance with the principles of Art. 2-3, the distributed loads have been transformed to their resultants P_1 and P_2, and these are represented by dashed vectors.

$$P_1 = 3 \times 4 = 12 \text{ kips} \qquad P_2 = 2 \times 11 = 22 \text{ kips}$$
$$\Sigma M_R = 15R_L - 12 \times 13 - 22 \times 5.5 = 0 \qquad \therefore R_L = \textbf{18.5 kips}$$
$$\Sigma M_L = 12 \times 2 + 22 \times 9.5 - 15R_R = 0 \qquad \therefore R_R = \textbf{15.5 kips}$$
$$\Sigma F = 18.5 + 15.5 - 12 - 22 = 0 \qquad \text{OK}$$

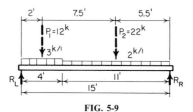

FIG. 5-9

PART *c*. (*Refer to Fig. 5-10.*)

$$P = \tfrac{1}{2}(7 + 3) \times 20 = 100 \text{ kips}$$

$$x = \frac{20}{3}\left(\frac{7 + 2 \times 3}{7 + 3}\right) = 8.67 \text{ ft}$$

$$\Sigma M_R = 20R_L - 100 \times 11.33 = 0 \qquad \therefore R_L = \textbf{56.7 kips}$$
$$\Sigma M_L = 100 \times 8.67 - 20R_R = 0 \qquad \therefore R_R = \textbf{43.3 kips}$$
$$\Sigma F = 56.7 + 43.3 - 100 = 0 \qquad \text{OK}$$

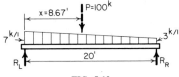

FIG. 5-10

PART *d*. (*Refer to Fig. 5-11.*)

$$P = 2 \times 9 = 18 \text{ kips}$$
$$\Sigma F = R - 5 - 18 = 0 \qquad\qquad \therefore \ R = \textbf{23 kips}$$
$$\Sigma M_R = M - 5 \times 9 - 18 \times 4.5 = 0 \qquad \therefore \ M = \textbf{126 ft-kips}$$
$$\Sigma M_L = 18 \times 4.5 - 23 \times 9 + 126 = 0 \qquad \text{OK}$$

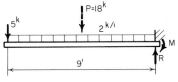

FIG. 5-11

PART *e*. (*Refer to Fig. 5-12.*)

$$P = 4 \times 21 = 84 \text{ kips}$$
$$\Sigma M_R = 15 R_L - 84 \times 10.5 + 10 \times 3 = 0 \qquad \therefore \ R_L = \textbf{56.8 kips}$$
$$\Sigma M_L = 84 \times 4.5 + 10 \times 18 - 15 R_R = 0 \qquad \therefore \ R_R = \textbf{37.2 kips}$$
$$\Sigma F = 56.8 + 37.2 - 84 - 10 = 0 \qquad \text{OK}$$

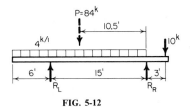

FIG. 5-12

It is frequently more convenient to compute the reactions for a simple beam by distributing the loads individually to the two supports. For example, in Fig. 5-8,

$$R_L = 3 \times \tfrac{10}{20} + 4 \times \tfrac{6}{20} = 2.7 \text{ kips}$$
$$R_R = 3 \times \tfrac{10}{20} + 4 \times \tfrac{14}{20} = 4.3 \text{ kips}$$

5-3 Vertical Shear and Bending Moment. Having evaluated the external forces acting on a beam, we now proceed to investigate the internal forces that are present. For this purpose, refer to the beam

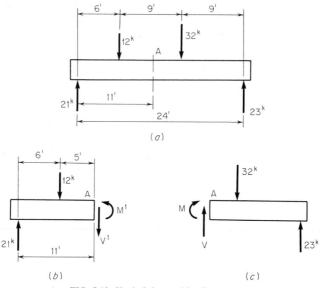

FIG. 5-13. **Vertical shear and bending moment.**

in Fig. 5-13a, and consider the member to be severed at section A. The parts of the beam to the left and to the right of this section are shown as free bodies in Fig. 5-13b and c, respectively. As these drawings reveal, both a vertical force and a moment are present at A. Force V' and moment M' in Fig. 5-13b constitute the action of the right part of the beam on the left part; force V and moment M in Fig. 5-13c constitute the action of the left part on the right part. Since bodies in contact exert opposing forces on one another, it follows that V and M are numerically equal to V' and M', respectively, but oppositely sensed. Applying the equations of equilibrium to the left part of the beam, we obtain

$$\Sigma F_V = 21 - 12 - V' = 0$$
$$\therefore V' = 21 - 12 = 9 \text{ kips} \qquad (a)$$
$$\Sigma M_A = 21 \times 11 - 12 \times 5 - M' = 0$$
$$\therefore M' = 21 \times 11 - 12 \times 5 = 171 \text{ ft-kips} \qquad (b)$$

The force V and moment M that the left part of the beam exerts on the right part are referred to as the *vertical shear* and *bending moment* at A, respectively. From Eqs. (a) and (b) we deduce the following:

V = algebraic sum of forces to left of given section

M = algebraic sum of moments of forces to left of given section with respect to centroidal axis of that section

For emphasis, we repeat the sign convention pertaining to the calculation of V and M: an external force is positive if it acts upward; the moment of an external force is positive if it is clockwise. Thus, the vertical shear and bending moment constitute a force system at the given section that is equivalent to the system of forces to the left of that section. It should be stressed that the symbol M denotes bending moment and the symbol ΣM denotes the algebraic sum of moments of *all* forces on a body with respect to some specified axis.

The vertical shear and bending moment that exist at each section of a beam can be represented graphically by constructing vertical-shear and bending-moment diagrams. In these diagrams, the abscissa represents the distance from the left end of the beam to the given section and the ordinate represents the vertical shear or bending moment at this section.

Example 5-2. Construct the vertical-shear and bending-moment diagrams for the beam in Fig. 5-14, and indicate the values at all significant sections.

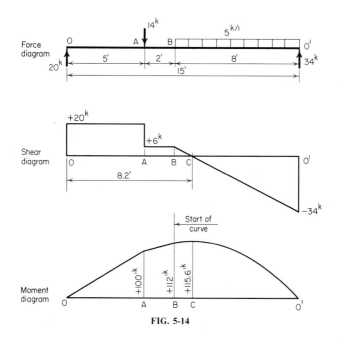

FIG. 5-14

Solution. Taking moments about the supports, we find that the reactions at the left and right supports, respectively, are

$$R_L = 20 \text{ kips} \qquad \text{and} \qquad R_R = 34 \text{ kips}$$

We shall formulate equations for vertical shear and bending moment, using subscripts to identify the boundary sections of the beam interval to which a given equation applies. Let x denote the distance from the left support to a given section.

$$V_{OA} = 20 \text{ kips} \qquad V_{AB} = 20 - 14 = 6 \text{ kips}$$
$$V_{BO'} = 6 - 5(x - 7) = 41 - 5x$$

V is zero when $x = 8.2$ ft.

$$M_{OA} = 20x$$
$$M_{AB} = 20x - 14(x - 5) = 6x + 70$$

In interval BO', the load from B to the given section is $5(x - 7)$, and the lever arm of this load is $\frac{1}{2}(x - 7)$. Then

$$M_{BO'} = 20x - 14(x - 5) - \frac{1}{2} \times 5(x - 7)^2$$
$$= -2.5x^2 + 41x - 52.5$$

The bending moment attains its maximum value in interval BO'. To locate the corresponding section, we set the derivative of bending moment equal to zero.

$$M = -2.5x^2 + 41x - 52.5$$

$$\frac{dM}{dx} = -5x + 41 = 0 \qquad \therefore x = 8.2 \text{ ft}$$

The maximum bending moment therefore occurs at C, where the vertical shear is zero. The values of bending moment at A, B, and C are recorded in the diagram.

Consider that the beam extends a short distance to the left of O and to the right of O'. From the definitions, it follows that the vertical shear and bending moment in these intervals are zero. Thus, the vertical-shear and bending-moment diagrams always close at their ends.

We shall now develop several properties that facilitate the construction of vertical-shear and bending-moment diagrams. Let A and B denote two beam sections an infinitesimal distance Δx apart. Figure 5-15 shows a part of the beam to the right of A. Let

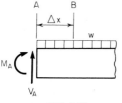

FIG. 5-15

w denote the pressure (load per unit of length, or unit load) in this interval. The vertical shear and bending moment at B can be expressed in terms of the values at A, giving

$$V_B = V_A - w\,\Delta x$$

$$M_B = M_A + V_A\,\Delta x - \frac{w(\Delta x)^2}{2}$$

Letting Δx approach zero and discarding subscripts, we obtain

$$\frac{dV}{dx} = -w \tag{5-1}$$

and

$$\frac{dM}{dx} = V \tag{5-2}$$

Then   Slope of vertical-shear diagram $= -w$
and   Slope of bending-moment diagram $= V$

Let ΔV denote the increment of vertical shear between two sections A and B, where B lies to the right of A. From Eq. (5-1),

$$\Delta V = -\int_A^B w\,dx \tag{5-3}$$

Similarly, let ΔM denote the increment of bending moment between these sections. From Eq. (5-2),

$$\Delta M = \int_A^B V\,dx \tag{5-4}$$

Interpreted geometrically, the increment of bending moment equals the area under the vertical-shear diagram bounded by sections A and B.

It follows as a corollary of Eq. (5-2) that the bending moment attains a local maximum or minimum value when the vertical shear is zero or passes through zero when it precipitously changes sign. From Eq. (5-2), we deduce the following additional properties of the bending-moment diagram:

1. The diagram is rectilinear in an interval where the vertical shear remains constant and curvilinear where the vertical shear varies (i.e., where a distributed load is present).

2. The diagram is concave downward in an interval where the vertical shear is diminishing. To rephrase this statement, the diagram has a center of curvature below the curve where a downward distributed load is present.

Example 5-3.   Construct the vertical-shear and bending-moment diagrams for the beam in Fig. 5-16.

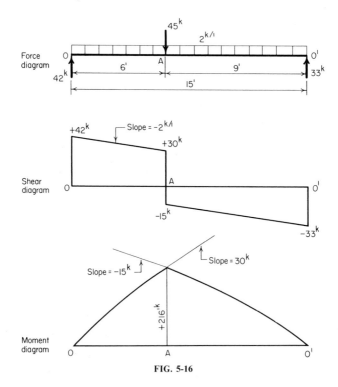

FIG. 5-16

Solution. Let $V_{A,L}$ and $V_{A,R}$ denote the vertical shear at an infinitesimal distance to the left and to the right of section A, respectively.

$$R_L = 42 \text{ kips} \qquad\qquad R_R = 33 \text{ kips}$$
$$V_{O,R} = 42 \text{ kips}$$
$$V_{A,L} = 42 - 2 \times 6 = 30 \text{ kips} \qquad V_{A,R} = 30 - 45 = -15 \text{ kips}$$
$$V_{O',L} = -15 - 2 \times 9 = -33 \text{ kips}$$
$$M_A = \text{area under } V \text{ diagram from } O \text{ to } A$$
$$\therefore M_A = \tfrac{1}{2}(42 + 30) \times 6 = 216 \text{ ft-kips}$$

Both branches of the bending-moment diagram are parabolic. The slope of each branch above point A is recorded in the diagram, although the problem does not require that this be done.

Example 5-4. Construct the vertical-shear and bending-moment diagrams for the overhanging beam in Fig. 5-17, and find the value of maximum positive and maximum negative bending moment.

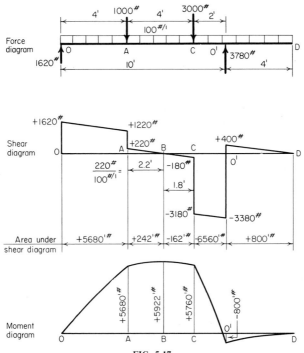

FIG. 5-17

Solution

$$\text{Total distributed load} = 1400 \text{ lb}$$
$$\Sigma M_R = 10R_L - 1400 \times 3 - 1000 \times 6 - 3000 \times 2 = 0$$
$$\therefore R_L = 1620 \text{ lb}$$
$$\Sigma M_L = 1400 \times 7 + 1000 \times 4 + 3000 \times 8 - 10R_R = 0$$
$$\therefore R_R = 3780 \text{ lb}$$

These values can be verified by the following calculation:

$$\Sigma F = 1620 + 3780 - 1000 - 3000 - 1400 = 0$$

The values of vertical shear at the significant sections and the areas under the diagram between these sections are recorded in the drawing.

$$M_A = 5680 \text{ ft-lb}$$
$$M_B = 5680 + 242 = 5922 \text{ ft-lb} = M_{max}$$
$$M_C = 5922 - 162 = 5760 \text{ ft-lb}$$
$$M_{O'} = 5760 - 6560 = -800 \text{ ft-lb} = M_{min}$$
$$M_D = -800 + 800 = 0$$

Example 5-5. Construct the vertical-shear and bending-moment diagrams for the beam in Fig. 5-18.

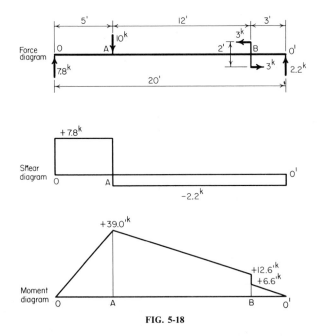

FIG. 5-18

Solution

$$\text{Moment of couple} = 3 \times 2 = 6 \text{ ft-kips}$$
$$\Sigma M_R = 20R_L - 10 \times 15 - 6 = 0 \qquad \therefore R_L = 7.8 \text{ kips}$$
$$\Sigma M_L = 10 \times 5 - 6 - 20R_R = 0 \qquad \therefore R_R = 2.2 \text{ kips}$$
$$M_A = 7.8 \times 5 = 39.0 \text{ ft-kips}$$
$$M_{B,L} = 39.0 - 2.2 \times 12 = 12.6 \text{ ft-kips}$$
$$M_{B,R} = 12.6 - 6 = 6.6 \text{ ft-kips}$$
$$M_C = 6.6 - 2.2 \times 3 = 0$$

The vertical-shear and bending-moment diagrams corresponding to various standard types of loading are presented in the AISC Manual. The reader should commit to memory the following equations for maximum bending moment:

For a simple beam with continuous uniform load,

$$M_{\max} = \tfrac{1}{8}wL^2 = \tfrac{1}{8}WL \tag{5-5}$$

where L = span
$\quad\; w$ = unit load
$\quad\; W$ = total load on beam = wL

For a simple beam with a concentrated load P at mid-span,

$$M_{\max} = \tfrac{1}{4}PL \tag{5-6}$$

5-4 Moving-Load Groups. Consider that a single concentrated load P rolls across a simple beam. Figure 5-19 shows the load at section A, and OBO' is the corresponding bending-moment diagram. As indicated, the maximum bending moment occurs at the section of the load, and its magnitude is as shown. Thus, associated with

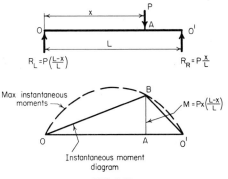

FIG. 5-19

each position of the load is an instantaneous bending-moment diagram, and the dashed parabola is the envelope of these instantaneous diagrams. The maximum moment induced during transit of the load (as distinguished from the *instantaneous* maximum moment) occurs when the load is at mid-span, and its magnitude is $\frac{1}{4}PL$. When the load is an infinitesimal distance from either support, the reaction at that support is P, and the absolute value of the maximum shear is therefore P.

Many structures, such as bridges and crane girders, support groups of concentrated loads that move across the structure while the loads remain fixed distances apart. It is therefore necessary to analyze the vertical shears and bending moments caused by such load groups. We shall consider first a group comprising two concentrated loads. In Fig. 5-20, let

P_1 = larger load and P_2 = smaller load
S = resultant of group = $P_1 + P_2$
a = distance between loads
x = distance from P_1 to the left support

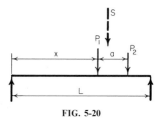

FIG. 5-20

Consider the interval of time when both loads are on the beam. It can be demonstrated that the maximum bending moment occurs when P_1 is as far from the left support as S is from the right support. Then

$$x = \frac{SL - P_2a}{2S} \tag{5-7}$$

and
$$M = \frac{(SL - P_2a)^2}{4SL} = \frac{Sx^2}{L} \tag{5-8}$$

However, if the distance a exceeds half the span, there is a possibility that the maximum moment induced by transit of the load group occurs when P_1 is at mid-span and P_2 lies to the right of the beam.

The maximum vertical shear occurs when P_1 lies an infinitesimal distance to the right of the left support. Then

$$V = P_1 + P_2 \frac{L - a}{L} \tag{5-9}$$

It should be stressed that we are concerned with the absolute rather than the algebraic value of the maximum vertical shear. In general, it is understood that the beam is simply supported.

Example 5-6.   A beam on a 40-ft span carries a moving-load group consisting of a concentrated load of 2 tons and one of 8 tons, with the loads spaced 14 ft apart. The loads must be increased by 30 percent to allow for impact. Determine the maximum bending moment and vertical shear caused by these loads.

Solution

$$P_1 = 16 \times 1.30 = 20.8 \text{ kips}$$
$$P_2 = 4 \times 1.30 = \underline{5.2}$$
$$S = 26.0 \text{ kips}$$

$$L = 40 \text{ ft} \qquad a = 14 \text{ ft}$$

$$M = \frac{(26.0 \times 40 - 5.2 \times 14)^2}{4 \times 26.0 \times 40} = \textbf{225 ft-kips}$$

$$V = 20.8 + 5.2 \times \tfrac{26}{40} = \textbf{24.2 kips}$$

Example 5-7.   Two concentrated loads of 20 and 5 kips move across a beam of 21-ft span while remaining 12 ft apart. Compute the maximum bending moment and vertical shear induced in the beam.

Solution

$$P_1 = 20 \text{ kips} \qquad P_2 = 5 \text{ kips} \qquad S = 25 \text{ kips}$$
$$L = 21 \text{ ft} \qquad a = 12 \text{ ft}$$

Since $a > L/2$, it is necessary to consider the two possible load positions inducing maximum moment. By Eq. (5-8), when both loads are on the beam, the maximum moment is

$$M = \frac{(25 \times 21 - 5 \times 12)^2}{4 \times 25 \times 21} = 103.0 \text{ ft-kips}$$

With the 20-kip load at mid-span, the maximum moment is

$$M = \tfrac{1}{4} \times 20 \times 21 = 105.0 \text{ ft-kips}$$
$$\therefore M_{\text{max}} = \textbf{105.0 ft-kips}$$

With the 20-kip load at the left support,

$$V = 20 + 5 \times \tfrac{9}{21} = \textbf{22.1 kips}$$

Example 5-8. The wheel base of a two-axle truck weighing 10 tons is 15 ft. If one-quarter of its weight is carred by the front axle, how shall the truck be placed on a 30-ft span to yield the maximum bending moment? How shall the truck be placed to yield maximum vertical shear?

Solution

$$P_1 = 15 \text{ kips} \qquad P_2 = 5 \text{ kips} \qquad S = 20 \text{ kips}$$
$$L = 30 \text{ ft} \qquad a = 15 \text{ ft}$$
$$x = \frac{20 \times 30 - 5 \times 15}{2 \times 20} = 13.1 \text{ ft}$$

Assume that the truck moves from right to left. The relative position of the loads is the reverse of that shown in Fig. 5-20. The bending moment is maximum when the rear axle is 13.1 ft from the *right* support; the vertical shear is maximum when the rear axle is an infinitesimal distance to the left of the right support.

When a moving-load group comprises more than two concentrations, the presence of many variable quantities precludes our formulating a universal rule for positioning the group to secure the maximum bending moment. The following trial-and-error procedure must be followed:

1. Locate the resultant of the load group.
2. Selecting a particular load, place the group in such position that this load is as far from one support as the resultant is from the other. If d denotes the distance from the load under consideration to the resultant, the distance from this load to the adjacent support is $\tfrac{1}{2}(L - d)$.
3. Compute the instantaneous moment at the section under this load. This is usually the greatest moment occurring under this load during transit of the load group.
4. Repeat this procedure for the remaining loads.

The third step is to be undertaken only if it is found, after positioning the loads in the prescribed manner, that all loads are present on the span and that the shear under this load is zero or passes through zero. Moreover, when the distances between loads are

large in relation to the span, the maximum moment induced by
transit of the load group may occur when only part of the group
is on the span.

For maximum shear, place all loads on the span, with the resultant
of the group at the minimum possible distance from a support.

Example 5-9. The load group shown in Fig. 5-21a moves across a simple
beam of 24-ft span while the distances between loads remain constant.

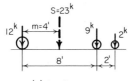

(a) Load group

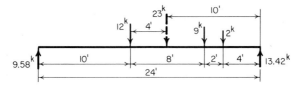

(b) First trial position for maximum moment

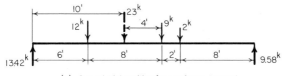

(c) Second trial position for maximum moment

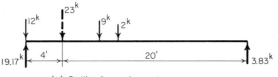

(d) Position for maximum shear

FIG. 5-21

Compute the maximum bending moment and vertical shear developed in the beam.

Solution

$$S = 12 + 9 + 2 = 23 \text{ kips}$$

To locate the resultant, take moments about the 12-kip load, giving

$$23m = 9 \times 8 + 2 \times 10 \qquad \therefore m = 4 \text{ ft}$$

For the 12-kip load,

$$d = 4 \text{ ft}$$
$$\tfrac{1}{2}(L - d) = \tfrac{1}{2}(24 - 4) = 10 \text{ ft}$$

Refer to Fig. 5-21*b*.

$$R_L = 23 \times \tfrac{10}{24} = 9.58 \text{ kips}$$
$$M \text{ at load} = 9.58 \times 10 = 95.8 \text{ ft-kips}$$

For the 9-kip load,

$$d = 4 \text{ ft} \qquad \tfrac{1}{2}(L - d) = 10 \text{ ft}$$

Refer to Fig. 5-21*c*.

$$R_L = 23 \times \tfrac{14}{24} = 13.42 \text{ kips}$$
$$M \text{ at load} = 13.42 \times 14 - 12 \times 8 = 91.9 \text{ ft-kips}$$

When the 2-kip load is positioned as directed, the shear at this load is not zero. Selecting the higher value, we have

$$M_{\max} = \mathbf{95.8 \text{ ft-kips}}$$

The vertical shear is maximum when the 12-kip load is an infinitesimal distance to the right of the left support. Refer to Fig. 5-21*d*.

$$V = R_L = 23 \times \tfrac{20}{24} = \mathbf{19.17 \text{ kips}}$$

If the maximum bending moment due to dead load is very small compared with that due to moving live loads, it is permissible to add the two bending moments, despite the fact that the sections at which these occur are generally not coincident.

5-5   Bending Stresses.   As loads are applied to a beam, the member deforms into a curved position, and we have found that internal moments are induced at the cross sections. These internal moments result from stresses, and we shall now analyze these stresses. For this purpose, we conceive the beam to be composed of fibers, a fiber being an infinitesimally narrow strip of material that runs along the length of the beam, parallel to its longitudinal axis. The stresses in these fibers that produce the bending moments are called *bending*, *flexural*, or *fiber* stresses.

The basic characteristic of the deformation of a beam under load is this: A cross section remains a plane as the beam deforms. Figure 5-22a shows a segment of a beam of length dx, bounded by cross sections A and B. The internal moments at the boundary sections are shown, but the internal shearing forces are omitted. To achieve greater clarity, the curvature of the beam is greatly exaggerated. It is seen that the internal moments at the boundary sections compress the upper fibers of the beam and distend the lower fibers, while the fibers at some intermediate surface, termed the *neutral*

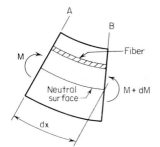

(*a*) Deformation of beam segment

(*b*) Bending stresses at A

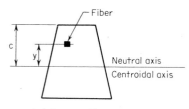

(*c*) Cross section

FIG. 5-22

surface, remain unstrained. Thus, bending stresses are *normal* stresses (tension or compression). Since the boundary sections are planes, the strain of a fiber is directly proportional to its distance from the neutral surface. Therefore, if the beam is homogeneous and the proportional limit is not exceeded, the bending stress in a fiber is also directly proportional to its distance from the neutral surface, as shown in Fig. 5-22b.

The line of intersection of the neutral surface and the cross section is called the *neutral axis*, as shown in Fig. 5-22c. Assume again that the beam was initially in a horizontal position. It can be demonstrated that the neutral axis coincides with the horizontal centroidal axis of the section. Let

M = bending moment at given cross section

I = moment of inertia of given section with respect to its horizontal centroidal axis

y = distance from neutral axis to given fiber

f = bending stress in given fiber at given section (in numerical value)

c = distance from neutral axis to outermost fiber

Then
$$f = \frac{My}{I} \qquad (5\text{-}10)$$

The maximum stress (in numerical value) occurs at the outermost fiber. Therefore,

$$f_{max} = \frac{Mc}{I} = \frac{M}{I/c} \qquad (5\text{-}11)$$

The quantity I/c is labeled the *section modulus* of the section and is represented by S. The unit of section modulus is in.^{3}. Equation (5-11) can therefore be expressed as

$$f_{max} = \frac{M}{S} \qquad (5\text{-}12)$$

Figure 5-22a reveals that if the bending moment at section A is positive, the bending stresses at that section are compressive in the fibers above the neutral axis and tensile in the fibers below. If the bending moment is negative, the stresses have the opposite character.

Example 5-10. A beam was formed by welding three steel plates having the dimensions shown in Fig. 5-23. If the maximum bending moment in the beam is 90 ft-kips, what is the maximum bending stress?

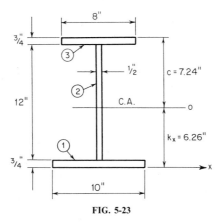

FIG. 5-23

Solution. The plates are numbered as indicated. The first step is to locate the centroidal axis of the section, and we shall do this by placing the x axis at the base and calculating the statical moment of the cross section with respect to this axis. Proceeding as in Example 2-2, we obtain the following results:

Segment	Area, sq in.	$\times$	k'_x, in.	$=$	Q'_x, in.^{3}
1	7.5		0.375		2.81
2	6.0		6.75		40.50
3	6.0		13.125		78.75
Total	19.5				122.06

Then
$$k_x = \frac{122.06}{19.5} = 6.26 \text{ in.}$$

Depth of section $= 0.75 + 12 + 0.75 = 13.50$ in.
$$c = 13.50 - 6.26 = 7.24 \text{ in.}$$

The next step is to calculate the moment of inertia of the section with respect to its centroidal axis. Refer again to Example 2-2.

Segment	$I_{o'}$, in.^{4}	k_o', in.	$A'k_o'^2$, in.^{4}
1	negligible	5.89	260.2
2	$\frac{1}{12} \times \frac{1}{2} \times 12^3 = 72.0$	0.49	1.4
3	negligible	6.87	283.2
Total	72.0		544.8

Then

$$I_o = 72.0 + 544.8 = 616.8 \text{ in.}^4$$
$$M = 90 \times 1000 \times 12 = 1,080,000 \text{ in.-lb}$$
$$f_{max} = \frac{Mc}{I} = \frac{1,080,000 \times 7.24}{616.8} = \textbf{12,680 psi}$$

Example 5-11. A beam having the cross section shown in Fig. 5-24*a* has the supports and loading shown in Fig. 5-24*b*. Find the maximum tensile and compressive bending stress in the member.

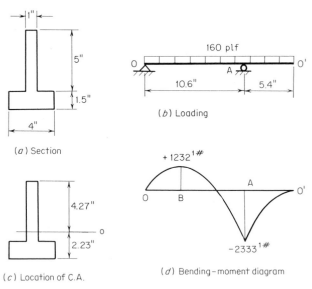

(*a*) Section

(*b*) Loading

(*c*) Location of C.A.

(*d*) Bending-moment diagram

FIG. 5-24

Solution. The location of the centroidal axis, which can be found most directly by applying Eq. (2-3), is shown in Fig. 5-24c. The moment of inertia of the section about this axis is

$$I_o = 40.3 \text{ in.}^4$$

As indicated in Fig. 5-24d, the extreme values of the bending moment are these:

$$M_B = 1232 \text{ ft-lb} \qquad \text{and} \qquad M_A = -2333 \text{ ft-lb}$$

The bending stresses at sections B and A are as follows:
At B:

$$f_{\text{top}} = \frac{1232 \times 12 \times 4.27}{40.3} = 1566 \text{ psi compression}$$

By proportion,

$$f_{\text{bottom}} = 1566 \times \frac{2.23}{4.27} = 818 \text{ psi tension}$$

At A:

$$f_{\text{top}} = 2966 \text{ psi tension} \qquad f_{\text{bottom}} = 1549 \text{ psi compression}$$

The extreme values of stress are therefore as follows:

$$\text{Maximum tensile stress} = \textbf{2966 psi}$$
$$\text{Maximum compressive stress} = \textbf{1566 psi}$$

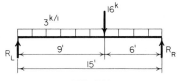

FIG. 5-25

Example 5-12. A W 18×50 steel beam, which has a section modulus of 89.1 in.^3, is loaded in the manner shown in Fig. 5-25. Compute the maximum stress due to flexure.

Solution

$$R_L = \tfrac{1}{2} \times 3 \times 15 + 16 \times \tfrac{6}{15} = 28.9 \text{ kips}$$

The vertical shear passes through a zero value at the concentrated load, and therefore the maximum bending moment occurs at that section.

$$M_{\text{max}} = 28.9 \times 9 - 3 \times 9 \times 4.5 = 138.6 \text{ ft-kips}$$
$$= 1,663,000 \text{ in.-lb}$$

By Eq. (5-12), $\qquad f_{\text{max}} = \dfrac{1,663,000}{89.1} = \textbf{18,700 psi}$

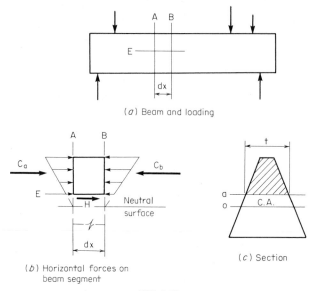

(a) Beam and loading

(b) Horizontal forces on beam segment

(c) Section

FIG. 5-26

5-6   Horizontal Shear and Shearing Stresses.   We shall now determine the manner in which the internal vertical shearing force is distributed across the transverse section of a beam. In Fig. 5-26a, A and B are two cross sections an infinitesimal distance dx apart, and E is a horizontal plane that we take for convenience as lying above the neutral surface. Consider that we extract from the beam the segment bounded by sections A and B and lying above plane E. Figure 5-26b shows the bending stresses acting on this segment, and C_a and C_b are the resultant bending forces on the segment. If the bending moment at B exceeds that at A, it follows that C_b exceeds C_a. Since the segment remains in equilibrium, it is seen that a horizontal force H exists at plane E; this is a force that the segment below E is exerting on the segment above E.

Let V denote the vertical shear at sections A and B. In Fig. 5-26c, o is the centroidal axis of the cross-sectional area and a is the line of intersection of plane E and the cross section. By applying Eqs. (5-2) and (5-10), we arrive at the following result:

$$H = \frac{VQ_{o,\text{above}}}{I} \, dx \qquad (5\text{-}13)$$

where $Q_{o,\text{above}}$ = statical moment of cross-sectional area above
line a with respect to o

I = moment of inertia of entire cross-sectional area
with respect to o

If sections A and B are a unit distance apart, the horizontal force acting on the segment is called the unit shearing force or *shear flow*. Let q denote the shear flow. Then

$$q = \frac{VQ_{o,\text{above}}}{I} \tag{5-14}$$

As shown in Fig. 5-26c, let t denote the width of the beam at line a. Assume that the horizontal shearing stress is uniform along line a, and let v denote this stress. Then

$$v = \frac{VQ_{o,\text{above}}}{It} \tag{5-15}$$

Theorem 3-1 in Art. 3-3 states that the shearing stresses on two mutually perpendicular planes are numerically equal. Therefore, Eq. (5-15) expresses the magnitude of the *vertical* shearing stress as well as the horizontal shearing stress at every point in the beam. This equation reveals the manner in which the vertical shear V is distributed across the transverse section.

Example 5-13. A beam having the cross section shown in Fig. 5-27a carries a load system that induces a maximum vertical shear of 2450 lb. Find the maximum shearing stress in the beam.

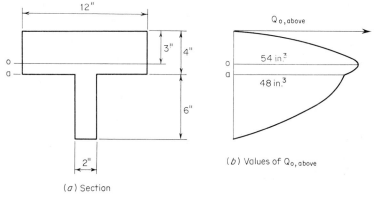

(a) Section

(b) Values of $Q_{o,\text{above}}$

FIG. 5-27

Solution. The location of the centroidal axis o is shown in Fig. 5-27a, and values of $Q_{o,above}$ are plotted in Fig. 5-27b. At o,

$$Q_{o,above} = 12 \times 3 \times 1.5 = 54 \text{ in.}^3$$

At a,
$$Q_{o,above} = 54 - 12 \times 1 \times 0.5 = 48 \text{ in.}^3$$

or, taking the entire area above a, we obtain

$$Q_{o,above} = 12 \times 4(3 - 2) = 48 \text{ in.}^3$$

At o,
$$\frac{Q_{o,above}}{t} = \frac{54}{12} = 4.5 \text{ in.}^2$$

At an infinitesimal distance below a,

$$\frac{Q_{o,above}}{t} = \frac{48}{2} = 24 \text{ in.}^2$$

The maximum shearing stress therefore occurs at a. The moment of inertia of the cross-sectional area about the centroidal axis is

$$I_o = 340 \text{ in.}^4$$

Applying Eq. (5-15), we obtain

$$v_{max} = \frac{2450 \times 24}{340} = \textbf{173 psi}$$

Consider that a beam has a rectangular cross section, and apply the notational system shown in Fig. 2-8. By symmetry, the horizontal centroidal axis lies at mid-depth, and the section modulus is

$$S = \frac{I}{c} = \frac{\dfrac{bd^3}{12}}{\dfrac{d}{2}}$$

or
$$S = \frac{bd^2}{6} \tag{5-16}$$

At the centroidal axis,

$$Q_{o,above} = b\frac{d}{2}\frac{d}{4} = \frac{bd^2}{8}$$

If V denotes the vertical shear at a given section and A denotes the cross-sectional area, the maximum shearing stress at the section is

$$v_{max} = \frac{VQ_{o,above}}{It} = \frac{V\dfrac{bd^2}{8}}{\dfrac{bd^3}{12}b} = \frac{1.5V}{bd}$$

or

$$v_{max} = \frac{1.5V}{A} \qquad (5\text{-}17)$$

Example 5-14. A beam supports a load system that induces a maximum bending moment of 26,000 ft-lb and a maximum vertical shear of 7800 lb. The beam is to have a rectangular section, and the allowable stresses are 1500 psi in bending and 120 psi in shear. Determine the theoretical dimensions of the section to the nearest hundredth of an in. (a) if the width is to be 0.75 times the depth; (b) if there is no restriction on the width-depth ratio.

Solution

$$M = 26,000 \times 12 = 312,000 \text{ in.-lb}$$

Applying Eqs. (5-12), (5-16), and (5-17), we obtain the following:

$$S = \frac{M}{f} = \frac{312,000}{1500} = 208 \text{ in.}^3 \qquad\qquad S = \frac{bd^2}{6}$$

$$\therefore bd^2 = 1248 \text{ in.}^3 \qquad (a)$$

$$A = \frac{1.5V}{v} = \frac{1.5 \times 7800}{120} = 97.5 \text{ sq in.} \qquad A = bd$$

$$\therefore bd = 97.5 \text{ sq in.} \qquad (b)$$

PART a: Replace b with $0.75d$ and solve for d.

From Eq. (a): $\qquad\qquad\qquad d = 11.85 \text{ in.}$
From Eq. (b): $\qquad\qquad\qquad d = 11.40 \text{ in.}$
$\qquad$ Set $d = \mathbf{11.85}$ **in.** $\qquad b = 0.75 \times 11.85 = \mathbf{8.89}$ **in.**

PART b: The objective is to minimize the weight of the member by providing the minimum area required. We shall apply the set of values that satisfies Eqs. (a) and (b) simultaneously. Divide the first equation by the second, giving

$$d = \frac{1248}{97.5} = \mathbf{12.80} \textbf{ in.} \qquad b = \frac{97.5}{12.80} = \mathbf{7.62} \textbf{ in.}$$

The presence of horizontal shearing forces in a beam can also be demonstrated intuitively, in this manner: Consider that the beam in Fig. 5-28a is severed along the horizontal plane MM, thereby dividing it into two parts. If friction is absent, these parts deform in the manner shown in Fig. 5-28b, a fiber directly above the cutting plane being elongated and a fiber directly below the plane being compressed. Thus, in every integral beam there is a tendency for each horizontal row of fibers to slide relative to its adjacent row, and this tendency can be resisted only by horizontal forces.

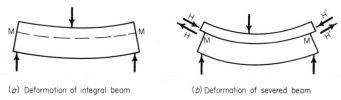

(*a*) Deformation of integral beam   (*b*) Deformation of severed beam

FIG. 5-28

If a beam is built up of parts, the connectors that join these parts must be capable of transmitting these horizontal forces in order to maintain the integrity of the member.

Figure 5-29 is the cross section of a typical wide-flange steel beam. The nomenclature for the section is as indicated.

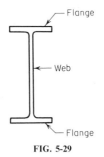

FIG. 5-29

Example 5-15.   A wide-flange beam was formed by welding together three 10- by $\frac{1}{2}$-in. plates of A36 steel. Determine the load this beam can carry at the center of a 10-ft span if the allowable fiber stress is 22,000 psi. Applying the AISC Specification, determine the disposition of $\frac{1}{4}$-in. fillet welds to connect the flange plates to the web plate.

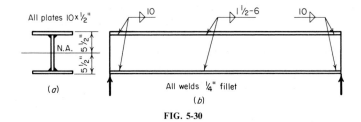

All plates 10 x ½"

N.A.

(a)

All welds ¼" fillet

(b)

FIG. 5-30

Solution. Refer to Fig. 5-30. Disregard the beam weight. By symmetry, the centroidal axis lies at mid-depth.

$$I = \tfrac{1}{12} \times \tfrac{1}{2} \times 10^3 + 2 \times 5 \times 5.25^2 = 317.3 \text{ in.}^4$$

$$M = \frac{PL}{4} = \frac{fI}{c}$$

$$\therefore P = \frac{4fI}{Lc} = \frac{4 \times 22,000 \times 317.3}{10 \times 12 \times 5.5} = \textbf{42,300 lb}$$

At the inner edge of the top flange,

$$Q_{o,\text{above}} = 5 \times 5.25 = 26.25 \text{ in.}^3$$

The shear flow at this surface is

$$q = \frac{VQ_{o,\text{above}}}{I} = \frac{\tfrac{1}{2} \times 42,300 \times 26.25}{317.3} = 1750 \text{ pli}$$

The allowable shearing stress on the weld throat is 21,000 psi, and the allowable shearing stress in the base metal is 14,500 psi. In accordance with the AISC Specification, the minimum length of intermittent weld is $1\tfrac{1}{2}$ in. and the weld spacing is restricted to 10.6 in. for the compression flange and 12 in. for the tension flange. The weld length at the supports is made equal to the width of member. The welds will be placed in pairs, one on each side of the web. Refer to Art. 4-5. For a 1-in. length of weld,

$$\text{Capacity of pair of welds} = 2 \times 0.707 \times \tfrac{1}{4} \times 1 \times 21,000$$
$$= 7420 \text{ pli}$$
$$\text{Capacity of plate} = \tfrac{1}{2} \times 1 \times 14,500 = 7250 \text{ pli}$$

Use a 1.5-in. length of weld; the capacity of the connection is $1.5 \times 7250 = 10,875$ lb.

Let h denote the center-to-center spacing of intermittent welds. Then

$$hq = \text{capacity of connection} \qquad \therefore h = \frac{10,875}{1750} = 6.2 \text{ in.}$$

For the intermittent welds, use pairs of $\frac{1}{4}$-in. welds, 1.5 in. long and 6 in. on centers, as shown in Fig. 5-30.

5-7 Composite Beams.   Figure 5-31 is the transverse section of a beam consisting of a timber core and steel side plates. We wish to compare the bending stresses in a fiber of steel and a fiber of timber, both lying at line *a*. Let the subscripts *s* and *t* refer to steel and timber, respectively.

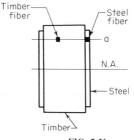

Timber fiber

Steel fiber

a

N.A.

Steel

Timber

FIG. 5-31

As the beam deforms, all fibers at line *a* undergo an identical deformation, regardless of the type of material. Let ϵ denote the strain of a fiber at line *a* of initial length dx. Then

$$\epsilon_s = \epsilon_t$$

Applying Eq. (3-6) but replacing *s* with *f*, we have

$$\frac{f_s}{E_s} = \frac{f_t}{E_t}$$

$$\therefore \frac{f_s}{f_t} = \frac{E_s}{E_t} = n \tag{5-18}$$

where *n* denotes the *modular ratio*, i.e., the ratio of the modulus of elasticity of steel to that of timber.

Example 5-16.   A timber beam having a width of 6 in. and depth of 8 in. is reinforced with two 6- by $\frac{1}{2}$-in. steel side plates, symmetrically disposed about mid-depth. If the maximum bending stress induced in the timber is 1300 psi, what is the maximum bending stress induced in the steel? Use $E_s = 30 \times 10^6$ psi and $E_t = 1.5 \times 10^6$ psi.

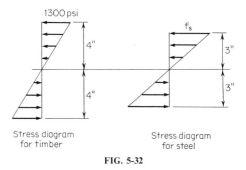

Stress diagram
for timber

Stress diagram
for steel

FIG. 5-32

Solution. Since the resultant tensile and compressive forces on a section are equal, the neutral axis lies at mid-depth by symmetry. As shown in Fig. 5-32, the maximum bending stress in timber and steel occurs 4 in. and 3 in., respectively, from the neutral axis.

$$n = \frac{E_s}{E_t} = \frac{30}{1.5} = 20$$

At a line 3 in. from the neutral axis,

$$f_t = 1300 \times \tfrac{3}{4} = 975 \text{ psi}$$
and
$$f_s = 20 \times 975 = \textbf{19,500 psi}$$

In analyzing the behavior of a composite beam, it is expedient to replace the given beam with a homogeneous beam that deforms in a manner identical to that of the given beam under a given load system. This equivalent homogeneous beam is termed the *transformed beam*, and its transverse section is termed the *transformed section*.

With reference to Fig. 5-31, consider that the steel fiber at line *a* is to be replaced with a timber fiber, also at line *a*. Let dA denote the cross-sectional area of the fiber, and let the subscript t' refer to the timber fiber that replaces the steel. Since the bending moment is not affected by this transformation, the bending force dF in the fiber remains constant. Then

$$dF = f_s \, dA_s = f_{t'} \, dA_{t'} \qquad \therefore \; dA_{t'} = \frac{f_s}{f_{t'}} \, dA_s$$

Thus, the transformed section is obtained by expanding the steel by the factor n parallel to the neutral axis and then replacing the steel with timber.

Example 5-17.   An 8- by 16-in. timber beam is reinforced with a 6- by $\frac{1}{2}$-in. steel plate at the bottom, as shown in Fig. 5-33a. If $E_s = 30 \times 10^6$ psi and $E_t = 1.2 \times 10^6$ psi, find the maximum bending stress in steel and timber when the bending moment is 480,000 in.-lb.

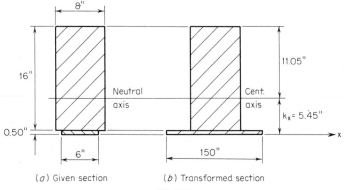

(*a*) Given section   (*b*) Transformed section

FIG. 5-33

Solution

$$n = \frac{30}{1.2} = 25 \qquad 6n = 150$$

The transformed section is shown in Fig. 5-33b.

$$A_1 = 8 \times 16 = 128 \text{ sq in.} \qquad A_2 = 0.50 \times 150 = 75 \text{ sq in.}$$

$$k_x = \frac{128 \times 8.50 + 75 \times 0.25}{128 + 75} = 5.45 \text{ in.}$$

$$I = \tfrac{1}{12} \times 8 \times 16^3 + 128 \times 3.05^2 + 75 \times 5.20^2 = 5949 \text{ in.}^4$$

$$f_{\text{top}} = \frac{480,000 \times 11.05}{5949} = 892 \text{ psi}$$

$$f_{\text{bottom}} = 892 \times 5.45/11.05 = 440 \text{ psi}$$

Therefore, the maximum stresses in the true beam are as follows:

In steel, $f_{\text{max}} = 25 \times 440 = \textbf{11,000 psi}$
In timber, $f_{\text{max}} = \textbf{892 psi}$

Example 5-18. With reference to the composite beam in Example 5-17, calculate the flexural capacity of the member if the allowable stresses are (a) 1400 psi in the timber and 20,000 psi in the steel; (b) 1600 psi in the timber and 18,000 psi in the steel. Assume that shearing stress is not critical.

Solution. Consider that the load on the member is gradually increased until one material attains its allowable stress. It is necessary to identify this material in order to calculate the allowable bending moment. From the results obtained in Example 5-17, we have

$$\frac{f_s}{f_t} = \frac{11{,}000}{892} = 12.3$$

PART a: Assume that $f_t = 1400$ psi. Then

$$f_s = 12.3 \times 1400 = 17{,}200 < f_{s,\text{allow}}$$

Thus, the timber limits the capacity. We therefore apply the properties of the transformed section with $f_{\text{top}} = 1400$ psi, giving

$$M = \frac{fI}{c} = \frac{1400 \times 5949}{11.05} = \textbf{754,000 in.-lb}$$

PART b: Assume that $f_t = 1600$ psi. Then

$$f_s = 12.3 \times 1600 = 19{,}700 > f_{s,\text{allow}}$$

Thus, the steel limits the capacity. Set $f_s = 18{,}000$ psi, giving

$$f_t = 18{,}000/12.3 = 1460 \text{ psi}$$

We therefore assign the value $f_{\text{top}} = 1460$ psi to the transformed section, and

$$M = \frac{1460 \times 5949}{11.05} = \textbf{786,000 in.-lb}$$

5-8 Combined Bending and Axial Loading. Many structural members function in the dual capacity of beams and tension or compression members. If the axial forces are tensile, we can combine the stresses induced by these forces with those caused by flexure; if the axial forces are compressive, we can also combine the stresses, provided that the axial forces do not cause the member to buckle.

Example 5-19. The concrete post shown in section in Fig. 5-34a carries a concentrated load of 300 kips applied at point Q. Compute the maximum and minimum stresses in the post.

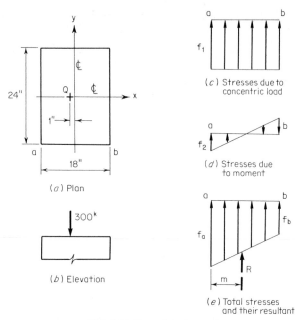

(a) Plan

(b) Elevation

(c) Stresses due to concentric load

(d) Stresses due to moment

(e) Total stresses and their resultant

FIG. 5-34. Eccentric load on post.

Solution. As stated in Art. 3-2, a load that acts at the centroid of a transverse section is *concentric* with respect to that section. The load in the present instance is said to be *eccentric*, and the distance from the centroid to the point at which the load acts is called the *eccentricity*. Applying the technique developed in Art. 1-3, we can transform the eccentric load to an equivalent force system consisting of a concentric load of the same magnitude and a moment about the y axis equal to the product of the load and its eccentricity. This moment causes compression to the left of the y axis and tension to the right. Let P denote the load, and refer to the stress diagrams in Fig. 5-34.

$$f_1 = \frac{P}{A} = \frac{300,000}{18 \times 24} = 694 \text{ psi}$$

$$f_2 = \frac{M}{S} = \frac{300,000 \times 1}{\frac{1}{6} \times 24 \times 18^2} = 231 \text{ psi}$$

$$f_a = 694 + 231 = \textbf{925 psi compression}$$
$$f_b = 694 - 231 = \textbf{463 psi compression}$$

These values can be tested by finding the resultant R of the stresses. It should be numerically equal to P and collinear with it. Applying the properties of the stress trapezoid in Fig. 5-34e, we have

$$R = \tfrac{1}{2} \times 18 \times 24(925 + 463) = 300{,}000 \text{ lb} \qquad \text{OK}$$

$$m = \frac{18}{3}\left(\frac{925 + 2 \times 463}{925 + 463}\right) = 8 \text{ in.} \qquad \text{OK}$$

With reference to Fig. 5-34, let e denote the eccentricity of the load and d the dimension of the rectangular section parallel to the direction of eccentricity. By setting $M = Pe$ and applying the formula for S, we obtain

$$f_a = \frac{P}{A}\left(1 + \frac{6e}{d}\right) \qquad \text{and} \qquad f_b = \frac{P}{A}\left(1 - \frac{6e}{d}\right) \tag{5-19}$$

Thus, applying the numerical data of Example 5-19, we have

$$f_a = \frac{300{,}000}{432}\left(1 + \frac{6 \times 1}{18}\right) = 694 \times 1.33 = 925 \text{ psi}$$

The equation for f_b reveals that this stress is positive if $e < d/6$. From this, it follows that the entire section has only one character of stress (tension or compression) if the load P is applied within the middle third of the x axis. If the load is applied outside this interval, both tension and compression exist within the member, provided that the member can withstand both types of stress. These relationships constitute the *principle of the middle third* for a rectangular section.

Now assume that a member of rectangular cross section is subjected to a load that is eccentric with respect to both center lines. Let

e_x and e_y = eccentricity of load with respect to x and y axes, respectively

d_x and d_y = dimension of section normal to x and y axes, respectively

An extension of Eqs. (5-19) yields the following:

$$f_{\max} = \frac{P}{A}\left(1 + \frac{6e_x}{d_x} + \frac{6e_y}{d_y}\right)$$

$$f_{\min} = \frac{P}{A}\left(1 - \frac{6e_x}{d_x} - \frac{6e_y}{d_y}\right) \tag{5-20}$$

Example 5-20. The post shown in Fig. 5-35a carries a concentrated load P applied at point Q. What is the maximum value of P if the stress in the post is restricted to 1000 psi?

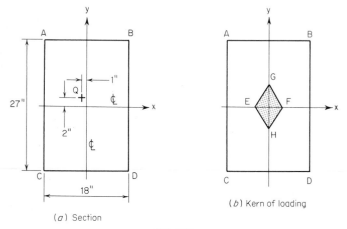

(a) Section

(b) Kern of loading

FIG. 5-35

Solution

$$A = 18 \times 27 = 486 \text{ sq in.} \qquad e_x = 2 \text{ in.} \qquad e_y = 1 \text{ in.}$$

Substituting in the first form of Eq. (5-20),

$$1000 = \frac{P}{486}\left(1 + \frac{6 \times 2}{27} + \frac{6 \times 1}{18}\right)$$

Solving, $P = \textbf{273,000 lb}$

The maximum stress occurs at A, and the minimum stress occurs at D.

In Fig. 5-35b, E, F, G, and H are the third points of their respective center lines. If the load P acts within the kern $EHFG$ or at its boundary, there is only one character of stress across the entire section; otherwise, both tension and compression exist. To illustrate the significance of the kern, assume that the post is made of a material that can withstand compression but not tension. A compressive load applied to the member must be confined within this kern.

PROBLEMS

5-1. Construct the shear and bending-moment diagrams for the beams in Fig. 5-36. Record the value of the shear and bending moment at every significant section.

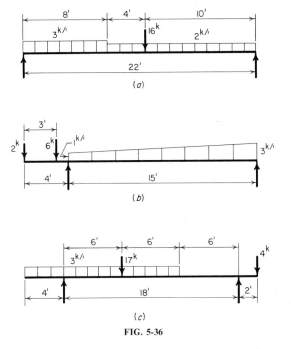

FIG. 5-36

5-2. With reference to the beam in Fig. 5-37, find the reactions at the supports and the maximum bending moment in the beam.

ANS. $R_L = 12$ kips; $R_R = 3$ kips; $M = 25$ ft-kips

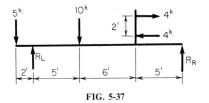

FIG. 5-37

5-3. The beam in Fig. 5-38 has a stationary support at the right, in the position indicated, and a movable support at the left. What is the minimum value of x that will prevent positive bending moment in the beam? ANS. 8 ft

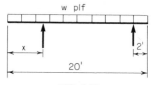

FIG. 5-38

5-4. Two concentrated loads of 16 kips and 12 kips move across a beam of 24-ft simple span while remaining 10 ft apart. Compute the maximum bending moment and maximum vertical shear induced in the beam.

ANS. 113.4 ft-kips; 23.0 kips

5-5. Two concentrated loads of 14 kips and 5 kips move across a beam of 36-ft simple span while remaining 21 ft apart. Compute the maximum bending moment in the beam.   ANS. 126 ft-kips

5-6. Two concentrated loads of 12 kips and 6 kips move across a beam of 15-ft simple span, and the maximum bending moment during transit of the system is 45 ft-kips. What is the minimum distance between the loads? Explain why an increase in this distance will not influence the maximum bending moment.   ANS. 8.25 ft

5-7. The axles of a three-axle truck are spaced 14 ft apart. The front axle carries 4 tons, the center axle 16 tons, and the rear axle 12 tons. What are the values of maximum bending moment and maximum shear as the truck moves across a simple span of 40 ft?   ANS. 420.9 ft-kips; 47.2 kips

5-8. With reference to the beam in Fig. 5-23, compute the maximum positive and maximum negative bending moments the beam can sustain if the allowable bending stresses in tension and compression, respectively, are as follows: (*a*) 18,000 and 15,000 psi; (*b*) 19,000 and 17,500 psi.

ANS. 1278 and −1478 in.-kips; 1491 and −1619 in.-kips

5-9. With reference to the beam in Fig. 5-39, find the following;

a. The location of the section between *A* and *B* at which the vertical shear is zero.   ANS. 6.12 ft from *A*

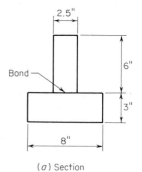

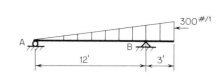

(*a*) Section   (*b*) Loading

FIG. 5-39

b. The maximum vertical shear in the beam (in absolute value).

ANS. 1065 lb

c. The maximum positive bending moment.   ANS. 1530 ft-lb

d. The maximum negative bending moment.   ANS. −1260 ft-lb

e. The maximum tensile bending stress.   ANS. 349 psi

f. The maximum compressive bending stress.   ANS. 424 psi

g. The maximum shearing stress on the bond.   ANS. 71 psi

5-10. The following data pertain to the beam shown in section in Fig. 5-40: At a section where the vertical shear is 3000 lb, the shear flow at line *a* is 200 pli. At a section where the bending moment is 204,000 in.-lb, the maximum flexural stress is 1060 psi. Find *b* and *d*.   ANS. $b = 7.39$ in.; $d = 12.50$ in.

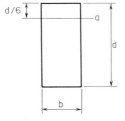

FIG. 5-40

5-11. A 4- by 8-in. timber beam (exact size) is reinforced by adding two vertical 6- by $\frac{1}{2}$-in. steel plates and one horizontal 8- by $\frac{1}{2}$-in. steel plate, in the manner shown in Fig. 5-41. The allowable bending stresses are 20,000 and 1500 psi, and the values of E are 30×10^6 and 1.5×10^6 psi for steel and timber, respectively. Compute the bending-moment capacity of the member.   ANS. 382,000 in.-lb

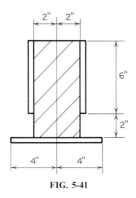

FIG. 5-41

5-12. The vertical member shown in plan in Fig. 5-42 carries a vertical compressive load applied at Q. The allowable stresses are 500 psi in compression and 100 psi in tension. What is the limiting value of the load at Q?   ANS. 39,300 lb

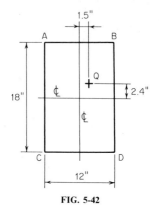

FIG. 5-42

5-13. With reference to Fig. 5-42, if a load applied to the member at Q induces a stress of 175 psi compression at A, what is the stress induced at C?

ANS. 92 psi tension

5-14. A member having a circular cross section with a diameter of 5 in. carries a load parallel to its longitudinal axis. The stress in the member caused by this load ranges from 1260 psi tension to 200 psi compression. Determine the magnitude of the load and its eccentricity. (The section modulus of a circle of diameter D is $\pi D^3/32$.)

ANS. 10,400 lb; 0.861 in.

6

DESIGN OF STEEL BEAMS
AND PLATE GIRDERS

6-1 Introduction. In the design of steel beams, we shall generally apply the Specification of the American Institute of Steel Construction.

As defined in Art. 3-4, the yield-point stress of structural steel is that stress at which a bar loaded in tension will undergo a large increase in strain without any increase in stress. The yield-point stress is denoted by F_y in the AISC Manual and by f_y in this text. Each allowable stress is a function of the yield-point stress, and Appendix A of the AISC Specification presents the set of values of allowable stress corresponding to each value of f_y. The most widely used grade of structural steel is A36, which has a yield-point stress of 36 ksi.

We shall confine our analysis of steel beams to the elastic-design method, which is described in Art. 3-8.

6-2 Distribution of Vertical Shear and Bending Moment. In order to design steel beams efficiently, it is necessary to appraise the relative contributions that the flanges and web make to the shearing and flexural strength of a rolled section.

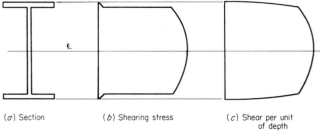

(*a*) Section   (*b*) Shearing stress   (*c*) Shear per unit
                                                              of depth

FIG. 6-1

Consider the steel beam shown in Fig. 6-1*a*. For simplicity, we are disregarding the fillet at juncture of flange and web. If we assign an arbitrary value to the vertical shearing force at a given section, calculate the shearing stress at every horizontal line, and plot these stresses, we obtain the diagram in Fig. 6-1*b*. If we now multiply the stress at each line by the corresponding width of section, we obtain the diagram of shear intensity, or shear per unit of depth, shown in Fig. 6-1*c*. Since this diagram approximates a rectangle, we may assume the following for simplicity:

1. The shear is resisted entirely by the web, provided that we equate the depth of web with the depth of section.

2. The shearing stress is uniform across the web.

To illustrate the application of these assumptions, consider the section W 16 × 36, which has a depth of 15.85 in. and web thickness of 0.299 in. Using an allowable shearing stress of 14,500 psi, we obtain the following value for the allowable vertical shear:

$$V_{\text{allow}} = 15.85 \times 0.299 \times 14,500 = 69,000 \text{ lb}$$

The values of allowable shear for each rolled-steel section corresponding to $f_y = 36$ ksi and $f_y = 50$ ksi are recorded at the bottom of the tables titled "Allowable uniform loads for beams laterally supported" in Part 2 of the AISC Manual.

For rolling purposes, a steel beam is generally provided with a web thickness that makes its shearing capacity very large in relation to its flexural capacity. As a result, it is usually the bending moment rather than the vertical shear that governs the selection of a steel beam. However, there are two situations in which the ratio of maximum vertical shear to maximum bending moment in a simply supported beam is relatively high: when the beam carries a heavy concentrated load near a support, and when the beam carries a distributed load on a short span. In these situations, after the section has been selected on the basis of its flexural capacity, its shearing capacity should be compared with the true vertical shear in the member.

Upon investigating some typical steel section, we find that the moment of inertia of its web is rather small in relation to that of its flanges. Consequently, the bending moment is resisted primarily by the flanges of the member.

In some instances, the design of a steel beam is controlled by the allowable deflection. However, we shall defer a study of beam deflections to a later point.

6-3   Allowable Bending Stress.   Consider that the slender member in Fig. 6-2 is subjected to concentric compressive forces applied at its ends. As these forces are gradually increased, a point is reached at which the member begins to buckle (deflect laterally), in the manner illustrated. Buckling introduces bending stresses that combine with the stresses due purely to axial loading, and failure of the member occurs when the axial forces are increased slightly beyond their value at incipient buckling. Since parts of a beam are in compression, potential buckling is an important consideration in beam design.

FIG. 6-2. Lateral deflection of compression member.

Failure of a steel beam by buckling can occur in three modes: buckling of the compression flange between points of lateral support, local buckling of the part of the compression flange that overhangs the web, and local buckling of the web. If the beam section is so proportioned as to exclude local buckling, the section is said to be *compact*. Those sections that are noncompact are identified by means of an asterisk in the allowable-uniform-load tables in Part 2 of the AISC Manual. For example, it is found that the section W 14 × 61 is noncompact when $f_y = 50$ ksi.

The allowable bending stress to be assigned to a beam is determined by the extent to which the member is prone to buckle. Let

f_b = allowable bending stress

L' = length of compression flange between points of lateral support

The range of values that L' can assume is divided into three intervals, and the boundary values of these intervals are denoted by L_c and L_u. These boundary values are recorded in the allowable-uniform-load tables in Part 2 of the AISC Manual. For example, with respect to the section W 24 × 94, the values are L_c = 9.6 ft and L_u = 15.1 ft when f_y = 36 ksi. The allowable bending stress associated with each interval of L' is given in Table 6-1 of this text. The value of f_b when the section is noncompact and $L' \leqslant L_c$ may be calculated by the AISC formula or it may be taken directly from the allowable-uniform-load tables in the AISC Manual. For example, with respect to the W 14 × 87, f_b = 31.5 ksi when $L' \leqslant L_c$ and f_y = 50 ksi.

TABLE 6-1.   Allowable Bending Stress

	Value of f_b	
Condition	Compact section	Noncompact section
$L' \leqslant L_c$	$0.66f_y$	Given in allowable-uniform-load tables
$L_c < L' \leqslant L_u$	$0.60f_y$	$0.60f_y$
$L' > L_u$	As determined by AISC equations	

In the following material, it is to be understood that the members are made of A36 steel. The selection of a beam size can be expedited by consulting the graphs of allowable bending moments appearing in Part 2 of the AISC Manual.

Example 6-1. A steel beam on a simple span of 20 ft carries a uniform load, including its estimated weight, of 1500 plf and a concentrated load of 4000 lb at a point 8 ft from the left support. The compression flange is unbraced along the entire span. Select the most economical beam size.

Solution. Refer to Fig. 6-3.

$$R_L = \tfrac{1}{2} \times 1.5 \times 20 + 4 \times \tfrac{12}{20} = 17.4 \text{ kips}$$

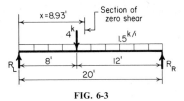

FIG. 6-3

At the section of zero shear,

$$V = 17.4 - 4 - 1.5x = 0 \qquad \therefore x = 8.93 \text{ ft}$$
$$M = 17.4 \times 8.93 - 4 \times 0.93 - \tfrac{1}{2} \times 1.5 \times 8.93^2 = 91.9 \text{ ft-kips}$$

We enter the allowable-bending-moment charts in Part 2 of the AISC Manual and locate the point having the abscissa $L' = 20$ ft and the ordinate $M = 91.9$ ft-kips. Proceeding vertically upward from this point until we reach the first solid line, we find that the W 12×45 is the lightest section that may be used. We shall verify the design. Since $L' = 20$ ft and $L_u = 17.8$ ft, it is necessary to apply the equations for f_b given in the AISC Specification.

The properties of the section are recorded in Part 1 of the AISC Manual, and the dimensions are shown in Fig. 6-4. The T section indicated in this drawing comprises the compression flange and one-sixth of the web area. The radius of gyration of this T section with respect to the center line of web is given as

$$r_T = 2.18 \text{ in.}$$

The ratio of the depth of member to the flange area is given as

$$\frac{d}{A_f} = 2.60 \text{ per in.}$$

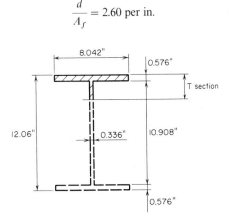

FIG. 6-4. Dimensions of W 12 × 45.

and the section modulus is

$$S = 58.2 \text{ in.}^3$$

Then
$$\frac{L'}{r_T} = \frac{20 \times 12}{2.18} = 110.1$$

and
$$\frac{L'd}{A_f} = 20 \times 12 \times 2.60 = 624$$

Refer to both the AISC Specification and Appendix A of the Specification. Since the maximum bending moment occurs within the unbraced interval rather than at a boundary section, $C_b = 1$. Since L'/r_T lies between the values 53 and 119, we apply the first equation containing this term, giving

$$f_b = 24.0 - \frac{110.1^2}{1181} = 13.7 \text{ ksi}$$

The equation containing d/A_f gives

$$f_b = \frac{12,000}{624} = 19.2 \text{ ksi}$$

The larger value of f_b governs. Then

$$S \text{ required} = \frac{M}{f_b} = \frac{91.9 \times 12}{19.2} = 57.4 < 58.2 \text{ in.}^3 \qquad \text{OK}$$

6-4   Cover-Plated Beams.   The flexural strength of a rolled section can be augmented by supplementing the beam with cover plates attached to the flanges. These reinforcing plates must span the region where the bending moment exceeds the capacity of the rolled section alone, and they must be extended beyond this region to develop their strength. The required size of cover plates can be obtained by applying the following approximation:

$$A = \frac{1.05(S - S_W)}{d_W} \qquad (6\text{-}1)$$

where A = area of cover plate
S = section modulus required
S_W = section modulus of W section
d_W = depth of W section

Example 6-2. A W 30 × 124 beam has beem fabricated, but a revision in the architectural plans requires that this member resist a bending moment of 954 ft-kips. Clearances permit the use of cover plates to be welded on the top and bottom to reinforce the member. The compression flange will have continuous lateral support. Determine the size of the cover plates.

Solution. Refer to Fig. 6-5. Assume that the reinforced section will be compact, thereby permitting a bending stress of 24 ksi.

$$S = \frac{M}{f} = \frac{954 \times 12}{24} = 477 \text{ in.}^3$$

$$I_W = 5360 \text{ in.}^4 \qquad S_W = 355 \text{ in.}^3 \qquad d_W = 30.16 \text{ in.}$$
$$\text{Flange width} = 10.52 \text{ in.}$$

Substituting in Eq. (6-1), we obtain

$$A = \frac{1.05(477 - 355)}{30.16} = 4.25 \text{ sq in.}$$

The minimum allowable distance from edge of cover plate to edge of rolled flange equals the weld size plus $\frac{5}{16}$ in. Try two plates 9 by $\frac{1}{2}$ in.

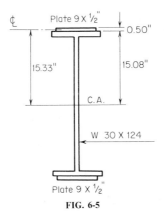

FIG. 6-5

The AISC Specification considers the reinforced section to be compact if the cover plate is welded continuously to the rolled section and the width-thickness ratio of the cover plate is restricted to $190/\sqrt{f_y}$, or 31.7. In the present instance, this ratio is 9/0.5, or 18.

Applying Eq. (2-9a), we obtain as the section modulus

$$S = \frac{I}{c} = \frac{5360 + 2 \times 9 \times 0.5 \times 15.33^2}{15.58} = 480 \text{ in.}^3 \qquad \text{OK}$$

In calculating the section modulus of a built-up riveted beam, the AISC Specification permits us to disregard the impairment of section caused by the rivet holes if the area of these holes does not exceed 15 percent of the gross flange area. Moreover, in the remaining cases, only the excess above 15 percent need be taken into account.

6-5   Design of Plate Girders.   A plate girder is a beam formed by the assemblage of simpler shapes, and it is used when the bending moment imposed on the member surpasses the flexural capacity of the strongest rolled section. As illustrated in Fig. 6-6, a riveted girder comprises a web plate and four flange angles, usually with the longer legs outstanding. The flange angles project $\frac{1}{4}$ in. beyond the edges of the web plate. Cover plates may be employed to supply up to 70 percent of the required flange area. In welded construction, horizontal plates are joined to the web plate at its top and bottom to form the flanges. The most economical depth of girder has been found to range between one-eighth and one-twelfth of the span, depending on the character of the loading.

FIG. 6-6. Riveted plate girder.

We shall apply the following notational system, part of which is illustrated in Fig. 6-7:

A_f = area of one flange
A_w = area of web
M = maximum bending moment

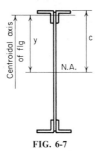

FIG. 6-7

y = distance from neutral axis to centroidal axis of flange
c = distance from neutral axis to extreme fiber
h = clear distance between flanges
t = thickness of web
f = flexural stress at extreme fiber

For simplicity, we shall assume tentatively that the entire flange area can be concentrated at its centroidal axis without modifying the resisting moment, and we shall assign to the web a depth of $2y$. Using this expedient, we obtain the following approximation:

$$M = \frac{2fy^2}{c}\left(A_f + \frac{A_w}{6}\right)$$

or

$$A_f = \frac{Mc}{2fy^2} - \frac{A_w}{6} \qquad (6\text{-}2)$$

The design of a plate girder is regulated by Sec. 1.10 of the AISC Specification. The procedure consists of the following steps:

1. Establish the depth of the member.
2. Determine the web thickness required to restrict the shearing stress to its allowable value or to conform to the AISC requirements.
3. Compute the required flange area by applying Eq. (6-2).
4. Select the flange angles and cover plates, if any.
5. Test the trial section.

Example 6-3. A plate girder of welded construction is to support the loads shown in Fig. 6-8a. The distributed load includes the weight of girder, and the top flange will have continuous lateral support. At its ends, the girder will bear on masonry buttresses. The total depth is restricted to approximately 70 in. Select the cross section.

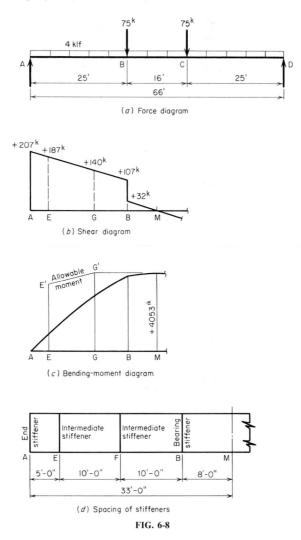

(a) Force diagram

(b) Shear diagram

(c) Bending-moment diagram

(d) Spacing of stiffeners

FIG. 6-8

Solution. The shear and bending-moment diagrams are shown in Fig. 6-8*b* and 6-8*c*, respectively.

We shall use a 68-in. web plate. Section 1.10.2 of the AISC Specification limits the slenderness ratio h/t to 322 in the absence of stiffeners having the stipulated spacing. However, if an allowable bending stress of 22 ksi is to be maintained, Sec. 1.10.6 imposes the following upper limit:

$$\frac{760}{\sqrt{22}} = 162$$

Then $\qquad t_{min} = \dfrac{h}{162} = \dfrac{68}{162} = 0.42$ in.

Use web plate 68 by $\frac{7}{16}$ in., giving

$$A_w = 29.75 \text{ sq in.}$$

By Sec. 1.5.1, the allowable shearing stress is 14.5 ksi. In the present instance,

$$v = \frac{V}{A_w} = \frac{207}{29.75} = 6.96 \text{ ksi} \qquad \text{OK}$$

Assume 1-in. flange plates. Equation (6-2) yields

$$A_f = \frac{4053 \times 12 \times 35}{2 \times 22 \times 34.5^2} - \frac{29.75}{6} = 27.54 \text{ sq in.}$$

Try 22- by $1\frac{1}{4}$-in. flange plates, giving

$$A_f = 27.5 \text{ sq in.}$$
Width-thickness ratio of projection $= 11/1.25 = 8.8$

The flange plates thus satisfy Sec. 1.9.1, which limits this ratio to 15.8.
Trial Section
   One web plate 68 by $\frac{7}{16}$ in.
   Two flange plates 22 by $1\frac{1}{4}$ in.
Applying Eqs. (2-12) and (2-9a), we have

$$I = \tfrac{1}{12} \times 0.438 \times 68^3 + 2 \times 27.5 \times 34.63^2 = 77,440 \text{ in.}^4$$

$$f = \frac{Mc}{I} = \frac{4053 \times 12 \times 35.25}{77,440} = 22.1 \text{ ksi} \qquad \text{OK}$$

The trial section is therefore satisfactory.

6-6   Bearing and Intermediate Stiffeners.

Since the sole function of the flanges of a steel beam is to provide bending-moment capacity, it is essential that any external forces that may be applied to the flanges be transmitted to the web by means of stiffeners. These bearing stiffeners are supplied in pairs, one on each side of the web; they must be milled to ensure intimate contact with the flange resisting the external force.

Example 6-4. With reference to the plate girder in Example 6-3, design the bearing stiffeners at the supports.

Solution. The reaction at each support is 207 kips, as shown in Fig. 6-8*b*. The design of bearing stiffeners is based on the provisions of Sec. 1.10.5.1 of the AISC Specification. The stiffeners are considered to act in conjunction with the tributary portion of the web to form a column section, as shown in Fig. 6-9. The area contributed by the web is

$$5.25 \times 0.438 = 2.30 \text{ sq in.}$$

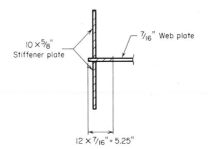

FIG. 6-9. Effective column section.

The stiffeners will be made 10 in. wide. By Sec. 1.9.1 of the Specification, the width-thickness ratio of the plates is restricted to $95/\sqrt{f_y}$, or 15.8. If $\frac{5}{8}$-in. plates are used, this ratio is 16, which may be considered acceptable.

The column capacity of the member formed by the stiffeners and tributary web area is found in this manner:

$$A = 2 \times 10 \times 0.625 + 2.30 = 14.80 \text{ sq in.}$$

Let I and r denote the moment of inertia and radius of gyration, respectively, of the section with respect to the center line of web.

$$I = \tfrac{1}{12} \times 0.625 \times 20.44^3 = 445 \text{ in.}^4$$

$$r = \sqrt{\frac{I}{A}} = \sqrt{\frac{445}{14.80}} = 5.48 \text{ in.}$$

The effective length KL of the member as a column is taken as three-fourths of the length of the stiffeners.

$$\frac{KL}{r} = \frac{0.75 \times 68}{5.48} = 9.3$$

Entering Table 1-36 in Appendix A of the AISC Specification with this slenderness ratio, we obtain an allowable axial compressive stress of 21.2 ksi. The true stress is

$$f = 207/14.80 = 14.0 \text{ ksi} \qquad \text{OK}$$

Section 1.5.1.5.1 provides an allowable bearing stress in the stiffeners of $0.90f_y$, or 33 ksi. In calculating the bearing area, assume that each stiffener will be clipped 1 in. to clear the flange-to-web welding. The true bearing stress is

$$f = \frac{207}{2 \times 9 \times 0.625} = 18.4 \text{ ksi} \qquad \text{OK}$$

The 10- by $\frac{5}{8}$-in. stiffeners at the supports are therefore satisfactory with respect to the width-thickness ratio, column action, and bearing.

Since the web of a plate girder is comparatively thin, failure can result from buckling of the web under compression along an inclined plane. To preclude this possibility, it is often necessary to provide web stiffeners at intermediate points along the span. The region of the beam between a successive set of stiffeners is termed a *panel*. The spacing of intermediate stiffeners is governed by Secs. 1.10.5.2 and 1.10.5.3 of the AISC Specification.

In accordance with Theorems 3-3 and 3-4 of Art. 3-3, the maximum compressive stress and maximum tensile stress at any point in the beam web occur on inclined planes that are mutually perpendicular. The tendency of the beam web to buckle under diagonal compression is partly resisted by diagonal tension; this behavior is known as *tension-field action*. The AISC Specification recognizes tension-field action except with respect to a hybrid (nonhomogeneous) girder, an end panel, and an interior panel containing a large hole.

Let a denote the panel length and, as before, let h denote the clear distance between flanges. The allowable value of the *aspect ratio* a/h depends on the shearing stress in the panel, the slenderness ratio h/t, and the presence or absence of tension-field action. Equations (1.10-1) and (1.10-2) of the Specification give allowable shearing stresses in terms of the values of a/h and h/t. The first equation ignores tension-field action, and the second takes this into account. The table "Plate Girders: Allowable shear stress in webs" in Part 2 of the AISC Manual provides a solution of Eq. (1.10-1) when $f_y = 36$ ksi, and Tables 3-36 to 3-100 in the Appendix of the Specification provide a solution of Eq. (1.10-2).

Example 6-5. With reference to the plate girder in Example 6-3, determine the spacing of intermediate stiffeners in the interval AB.

Solution. The stiffener spacing is shown in Fig. 6-8d. At the support, the shearing stress is 6.96 ksi.

$$\frac{h}{t} = \frac{68}{0.438} = 155 \qquad \left(\frac{260}{155}\right)^2 = 2.81$$

Referring to the table in Part 2 of the AISC Manual, we find by linear interpolation that a/h for the end panel is restricted to 0.89. Then

$$a = 0.89 \times 68 = 60.5 \text{ in.}$$

∴ Provide stiffeners at 5 ft from the ends.

In analyzing interval EB with respect to stiffeners, we may avail ourselves of tension-field action. At E,

$$V = 207 - 5 \times 4 = 187 \text{ kips} \qquad v = \frac{187}{29.75} = 6.29 \text{ ksi}$$

If stiffeners are not required in EB, we have

$$a = \tfrac{1}{2} \times 12(66 - 16) - 60 = 240 \text{ in.}$$

$$\frac{a}{h} = \frac{240}{68} > 3$$

Stiffeners are therefore required in EB, in accordance with Sec. 1.10.5.3. Assume tentatively that stiffeners may be placed at F, the center of EB. Then

$$\frac{a}{h} = \frac{120}{68} = 1.76 < 2.81 \qquad \text{OK}$$

From Table 3-36,

$$v_{allow} = 7.83 > 6.29 \text{ ksi} \qquad \text{OK}$$

It now remains to investigate the following with respect to interval EB: the combined shearing and bending stress, and the bearing stress. Equation (1.10-7) in effect reduces the allowable bending moment wherever the vertical shear exceeds 0.6 times the allowable shear.

$$V_{allow} = A_w v_{allow} = 29.75 \times 7.83 = 233 \text{ kips}$$
$$0.6 \times 233 = 140 \text{ kips}$$

In Fig. 6-8*b*, *G* denotes the section at which $V = 140$ kips. The allowable bending moment must be reduced to the left of *G*.

$$AG = (207 - 140)/4 = 16.75 \text{ ft}$$

At *G*,
$$M_G = 2906 \text{ ft-kips} \qquad M_E = 985 \text{ ft-kips}$$
$$M_{\text{allow}} = 4053 \text{ ft-kips}$$

At *E*,
$$f_{\text{allow}} = 36(0.825 - 0.375 \times 187/233) = 18.9 \text{ ksi}$$

$$M_{\text{allow}} = \frac{18.9 \times 77{,}440}{35.25 \times 12} = 3460 \text{ ft-kips}$$

In Fig. 6-8*c*, points *E′* and *G′* represent these allowable moments. Since Eq. (1.10-7) is linear and the shear in the beam varies linearly, we may connect points *E′* and *G′* with a straight line to obtain the allowable moments at intermediate sections. The true moment is below the allowable moment throughout interval *EB*.

Assume that the distributed load of 4 klf bears on the top flange of the member and thereby restrains that flange against rotation. Equation (1.10-10) of the Specification yields an allowable bearing stress in the web of 2.81 ksi. The true bearing stress is

$$\frac{4}{12 \times 0.438} = 0.76 \text{ ksi} \qquad \text{OK}$$

The stiffener spacing in interval *EB* is therefore satisfactory in all respects.

6-7 Rivet Pitch in Plate Girders.

The component parts of a built-up beam must be adequately stitched together to prevent the separation of these parts as a result of the horizontal shearing forces.

Let *p* denote the pitch of the flange-to-web rivets of a plate girder. Figure 6-10 shows a segment of the girder *p* in. in length, with C_a and

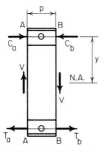

FIG. 6-10. Rivet pitch.

C_b representing the resultant flexural forces in the top flange at their respective sections. Assume for simplicity that the following are coincident: the action lines of C_a and C_b, the centroidal axis of the flange, and the gage line of the rivets. Let A denote the flange area and Q the statical moment of this area about the centroidal axis. Then

$$Q = Ay$$

The rivet connecting the compression flange to the web exerts a force R on the flange equal to the difference between C_a and C_b. Applying Eq. (5-14) for shear flow, we obtain

$$R = C_b - C_a = \frac{VQp}{I}$$

or

$$R = \frac{VAyp}{I} \tag{6-3}$$

By equating R with the rivet capacity, we can obtain the allowable pitch. However, if a distributed load bears on the top flange, it induces a vertical force on the rivet, and its effect must be taken into account.

Example 6-6. A plate girder is composed of the following material:
One web plate 48 by $\frac{3}{8}$ in.
Four flange angles 6 × 4 × $\frac{3}{4}$ in., with the long legs outstanding
Two cover plates 14 by $\frac{1}{2}$ in.
The overall depth of the girder is 49.5 in. Each flange is connected to the web plate by a single row of $\frac{7}{8}$-in. rivets spaced 3 in. on centers and located $2\frac{1}{2}$ in. from the back of angles. The compression flange has continuous lateral support, but there is no distributed load bearing on the top flange. Determine the flexural capacity of the girder and its shear capacity as governed by the rivet spacing.

Solution. Refer to Fig. 6-11. The capacity of a $\frac{7}{8}$-in. rivet bearing on a $\frac{3}{8}$-in. plate is 15.9 kips. The moment of inertia of the section about its centroidal axis is found by applying Eq. (2-9a). For the two flange angles:

$$A = 13.88 \text{ sq in.} \qquad I_{o'} = 17.4 \text{ in.}^4$$

The distance from the centroidal axis of the section to the centroidal axis of the angles is

$$k' = 24.25 - 1.08 = 23.17 \text{ in.}$$
$$\text{Gross flange area} = 13.88 + 7.0 = 20.88 \text{ sq in.}$$

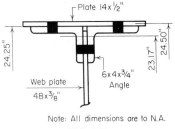

Note: All dimensions are to N.A.

FIG. 6-11

Taking the effective diameter of rivet holes as 1 in., we have

$$\text{Area of holes} = 2 \times \tfrac{1}{2} \times 1 + 4 \times \tfrac{3}{4} \times 1 = 4.00 \text{ sq in.}$$
$$\text{Allowable area} = 0.15 \times 20.88 = 3.13$$
$$\text{Excess area} = \overline{0.87} \text{ sq in.}$$

We shall consider this excess area to be extracted from the outstanding legs of the flange angles; its k' distance is therefore 23.88 in.

I of section

One web plate, $I_{o'}$	3,456 in.^{4}
Four flange angles, $I_{o'}$	35
$Ak'^2 = 2 \times 13.88 \times 23.17^2$	14,900
Two cover plates:	
$Ak'^2 = 2 \times 7.0 \times 24.50^2$	8,400
I of gross section	26,791
Deduct $2 \times 0.87 \times 23.88^2$	991
I of net section	25,800 in.^{4}

$$Ak' \text{ of flange} = 13.88 \times 23.17 + 7.0 \times 24.50 - 0.87 \times 23.88$$
$$= 472 \text{ in.}^3$$

This value will be applied for Ay in Eq. (6-3). For the web,

$$\frac{h}{t} = \frac{48.5 - 8}{0.375} = 108$$

Since the limit imposed by Sec. 1.10.6 of the AISC Specification is 162, the allowable flexural stress remains 22 ksi.

$$M = \frac{fI}{c} = \frac{22 \times 25{,}800}{24.75 \times 12} = \textbf{1911 ft-kips}$$

$$V = \frac{RI}{Ayp} = \frac{15.9 \times 25{,}800}{472 \times 3} = \textbf{290 kips}$$

7

DEFLECTION OF BEAMS—
STATICALLY INDETERMINATE BEAMS

7-1 Definitions and Notation.   We shall now analyze the manner in which a beam deforms under a given load system. For convenience, consider the beam to be weightless and assume that it lies in a horizontal position prior to loading. All cross sections are therefore vertical, and the longitudinal axis of the member, which is a line passing through the centroid of every cross section, is a horizontal straight line.

As the load system is applied, every cross section undergoes a translation and a rotation about its horizontal centroidal axis, thereby causing the longitudinal axis to bow into a new position that is called the *elastic curve* of the beam. Since the curvature of the elastic curve is generally extremely small, we may equate distances along this curve to their horizontal projections. The vertical displacement of the longitudinal axis at a given point along the beam as load is applied is called the *deflection* of the beam at that point.

Figure 7-1*a* shows a simply supported beam in its deformed position, the deformation being shown in hyperbolic fashion to accentuate its characteristics. The origin of coordinates may be placed at any convenient point on the original longitudinal axis. Let

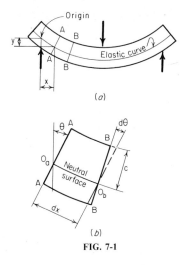

FIG. 7-1

x and y denote the horizontal and vertical coordinates, respectively, of a point on the elastic curve, and let θ denote the angle (in radians) between the tangent to the elastic curve at that point and the horizontal. Since the elastic curve is normal to every cross section, θ also equals the angle between the cross section at the given point and a vertical plane. The sign convention is as follows: x is positive if the given point lies to the *right* of the origin; y is positive if the displacement is *downward*; θ is positive if the tangent inclines *downward to the right* or, equivalently, if the cross section rotates in a clockwise direction.

Since θ is very small, we may equate this angle to its tangent function, giving

$$\theta = \frac{dy}{dx} \tag{7-1}$$

7-2 Curvature of Elastic Curve.   We shall now develop the basic relationship pertaining to beam deformations. In doing so, we shall disregard the deformation due to shear, as this is usually negligible. With reference to the beam in Fig. 7-1a, consider the beam segment of length dx bounded by sections A and B. This segment is shown in isolation in Fig. 7-1b. The rotation of B relative to A is numerically equal to the difference between their rotations, which we denote $d\theta$. Let

$M =$ bending moment at A

$E =$ modulus of elasticity of material

$I =$ moment of inertia of cross-sectional area with respect to its centroidal axis

$K =$ curvature of elastic curve at A (in absolute value)

By combining Eqs. (3-6) and (5-10), we obtain this result:

$$\frac{d\theta}{dx} = -\frac{M}{EI} \qquad (7\text{-}2)$$

In mathematics, the *curvature* of a curve is defined as the rate of change of the inclination of its tangent with respect to distance along the curve. Since in the present instance we are equating the distance along the curve to its horizontal projection, we may interpret Eq. (7-2) as an expression of curvature, giving

$$K = \frac{M}{EI} \qquad (7\text{-}3)$$

This relationship is seen to be logical because it states that the curvature is directly proportional to M (which creates the curvature), inversely proportional to E (which measures the stiffness of the material), and inversely proportional to I (which measures the dispersion of the material from its centroidal axis, and therefore its resistance to rotation). Thus, we may regard the product EI as the resistance that the beam offers to bending.

If the bending moment remains constant across an interval of a prismatic beam, the curvature of the elastic curve remains constant, and the elastic curve is therefore a circular arc. Because the curvature of a circular arc is the reciprocal of its radius R, Eq. (7-3) yields

$$R = \frac{EI}{M} \qquad (7\text{-}4)$$

Example 7-1. A steel bar 3 in. wide and $\frac{1}{2}$ in. thick is to be bent into a circular arc of 60-ft radius. What moment must be applied at its ends? Use $E = 30 \times 10^6$ psi.

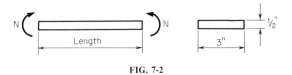

FIG. 7-2

Solution. Refer to Fig. 7-2. The bending moment throughout the bar equals the moment N applied at the ends.

$$I = \frac{3(\frac{1}{2})^3}{12} = \frac{1}{32} \text{ in.}^4$$

By Eq. (7-4),

$$N = \frac{EI}{R} = \frac{30 \times 10^6 \times \frac{1}{32}}{60 \times 12} = \textbf{1300 in.-lb}$$

If the bending moment in a given interval of a beam is positive, there is compression above the neutral surface and tension below it. Consequently, the elastic curve is concave upward (i.e., the center of curvature lies above the curve). On the other hand, if the bending moment is negative, the elastic curve is concave downward. If the bending moment is zero at a single section, the elastic curve has a point of inflection at that section. Finally, if the bending moment is zero across an interval, the corresponding portion of the elastic curve is a straight line.

7-3   Double-Integration Method.   Numerous techniques for analyzing the deformation of a beam have been devised, but we shall confine our study to the double-integration method. According to Eqs. (7-1) and (7-2),

$$\theta = \frac{dy}{dx} \qquad \text{and} \qquad \frac{d\theta}{dx} = -\frac{M}{EI}$$

$$\therefore \frac{d^2y}{dx^2} = -\frac{M}{EI} \tag{7-5}$$

By integrating the expression at the right twice and establishing the values of the two constants of integration that arise, we obtain the equation of the elastic curve.

Example 7-2. A prismatic cantilever beam of length L carries a load P applied at its free end. Derive the equation of the elastic curve. Find the slope of this curve at the free end.

   Solution.   Orient the beam in the manner shown in Fig. 7-3a, and place the origin at the free end. As seen in Fig. 7-3b, the bending moment at a given section is $-Px$. Since E and I are constant, they may be placed at the left side of Eq. (7-5). Then

$$EI \frac{d^2 y}{dx^2} = Px$$

$$EI \frac{dy}{dx} = \frac{Px^2}{2} + c_1 \tag{a}$$

$$EIy = \frac{Px^3}{6} + c_1 x + c_2 \tag{b}$$

The known boundary conditions are as follows: When $x = L$, $dy/dx = 0$ and $y = 0$. Substituting these values in Eqs. (a) and (b), we obtain

$$c_1 = -\frac{PL^2}{2} \quad \text{and} \quad c_2 = \frac{PL^3}{3}$$

Then
$$\frac{dy}{dx} = \frac{P}{2EI}(x^2 - L^2) \tag{a'}$$

and
$$y = \frac{P}{6EI}(x^3 - 3L^2 x + 2L^3) \tag{b'}$$

When $x = 0$,
$$\frac{dy}{dx} = -\frac{PL^2}{2EI}$$

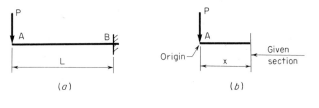

(a)

(b)

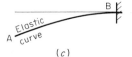

(c)

FIG. 7-3

Example 7-3. A prismatic cantilever beam of length L carries a distributed load that varies uniformly from zero at the free end to u plf at the fixed end. Derive the equation of the elastic curve in terms of the total load W on the beam.

Solution. Refer to Fig. 7-4. Let u_x denote the pressure at a given section. By proportion,

$$\frac{u_x}{u} = \frac{x}{L} \qquad \therefore u_x = \frac{ux}{L}$$

$$\text{Load to left of given section} = \frac{ux^2}{2L}$$

$$M = -\frac{ux^2}{2L}\frac{x}{3} = -\frac{ux^3}{6L}$$

Then

$$EI\frac{d^2y}{dx^2} = \frac{ux^3}{6L}$$

$$EI\frac{dy}{dx} = \frac{ux^4}{24L} + c_1 \tag{c}$$

$$EIy = \frac{ux^5}{120L} + c_1x + c_2 \tag{d}$$

Applying the identical boundary conditions as in Example 7-2, we obtain

$$c_1 = -\frac{uL^3}{24} \qquad \text{and} \qquad c_2 = \frac{uL^4}{30}$$

Then

$$y = \frac{u}{120EIL}(x^5 - 5L^4x + 4L^5) \tag{d'}$$

Now,

$$W = \frac{uL}{2} \qquad \therefore u = \frac{2W}{L}$$

Substituting in Eq. (d'),

$$y = \frac{W}{60EIL^2}(x^5 - 5L^4x + 4L^5)$$

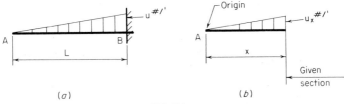

(a) (b)

FIG. 7-4

7-4   Calculation of Deflection by Superposition.   Many   reference books, including the Manual of the American Institute of Steel Construction, record the equation of the elastic curve of a beam associated with certain standard types of load systems. If the given system is a composite of two or more standard types, the equation of the elastic curve may be formed merely by combining the appropriate equations.

Example 7-4.   The overhanging beam in Fig. 7-5 has the properties $I = 400$ in.[4] and $E = 30 \times 10^6$ psi. Compute the deflection at C, using the following values: $w = 2$ klf, $P = 3$ kips, $L = 12$ ft, and $a = 4$ ft.

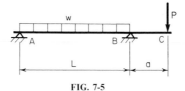

FIG. 7-5

Solution.   This load system is a composite of the following standard systems: a uniformly distributed load from A to B, and a concentrated load at C. Refer to Sec. 2 of the AISC Manual. By combining the two equations presented in the Manual but using the proper algebraic signs, we obtain the following:

$$y_C = \frac{Pa^2(L + a)}{3EI} - \frac{wL^3a}{24EI} = \frac{8Pa^2(L + a) - wL^3a}{24EI}$$

In substituting numerical values in an elastic-curve equation, it is convenient to express all values in the numerator in units of lb and ft and then to multiply by 1728 to convert cu ft to cu in. Then

$$y_C = \frac{(8 \times 3000 \times 4^2 \times 16 - 2000 \times 12^3 \times 4) \times 1728}{24 \times 30 \times 10^6 \times 400}$$

$$= -0.046 \text{ in.}$$

The negative result signifies that the free end is elevated.

7-5   Relative Rigidity of Beams   When a given load system is applied to a beam, the deformation at a given section is inversely proportional to the product EI, and the flexural stress is inversely proportional to the section modulus S. Therefore, we may regard EI as an index of the rigidity of the beam and S as an index of its flexural strength, provided that buckling of the compression flange is not a criterion.

Example 7-5. Two W 8 × 28 beams may be placed side by side or they may be riveted together to form a 16-in. beam, as shown in Fig. 7-6. How will the flexural stresses and deflections corresponding to the two arrangements compare? The properties of a single beam are as follows:

$$A = 8.23 \text{ sq in.} \qquad d = 8.06 \text{ in.} \qquad I = 97.8 \text{ in.}^4 \qquad S = 24.3 \text{ in.}^3$$

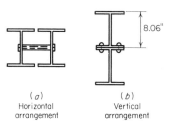

(*a*)
Horizontal
arrangement

(*b*)
Vertical
arrangement

8.06"

FIG. 7-6. Compound beam.

Solution.   Let the subscripts h and v pertain to the horizontal and vertical arrangements, respectively.

$$I_h = 2I = 195.6 \text{ in.}^4 \qquad S_h = 2S = 48.6 \text{ in.}^3$$
$$I_v = 2(I + Ak^2) = 2(97.8 + 8.23 \times 4.03^2) = 462.9 \text{ in.}^4$$

$$S_v = \frac{I_v}{8.06} = 57.4 \text{ in.}^3$$

Then
$$\frac{f_h}{f_v} = \frac{S_v}{S_h} = \frac{57.4}{48.6} = \mathbf{1.18}$$

and
$$\frac{y_h}{y_v} = \frac{I_v}{I_h} = \frac{462.9}{195.6} = \mathbf{2.37}$$

7-6   Analysis of Beams with Three Reactions.   The force system acting on a beam normally consists of vertical forces and couples, and therefore the system must satisfy two equations of equilibrium: $\Sigma F_v = 0$ and $\Sigma M = 0$. It follows that a beam having more than two reactions is statically indeterminate.

If a beam has three reactions, it is usually susceptible to analysis by the simple device of considering one reaction to be an active rather than a passive force or couple and then applying the known characteristics of the elastic curve of the member. The following example illustrates the procedure.

Example 7-6.   The supports for a beam consist of two simple supports lying 36 ft apart and at the same level and a third support located midway between the end supports in the horizontal direction and $1\frac{1}{4}$ in. below them. The beam is a W 21 × 73 and it carries a load of 3800 plf between the end supports. Compute the reaction and bending moment at the center support and the maximum bending stress in the beam. The properties of the section are $I = 1600$ in.^{4} and $S = 151$ in.^{3}, and $E = 30 \times 10^6$ psi.

Solution.   Refer to Fig. 7-7. We shall assume that the designated load includes the beam weight. The total load W on the beam is

$$W = 36 \times 3800 = 136,800 \text{ lb}$$

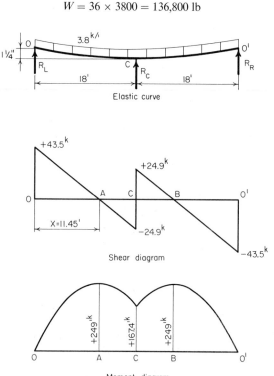

Elastic curve

Shear diagram

Moment diagram

FIG. 7-7. Deflection of beam to center support.

Remove the center support and treat R_C as a load rather than a reaction. In this manner, the beam is transformed to a simple beam that sustains a uniformly distributed load and a concentrated upward load at mid-span. According to the data for the elastic curve of a simple beam presented in

Sec. 2 of the AISC Manual, the deflections at C caused by these loads are as follows:

$$y_1 = \frac{5WL^3}{384EI} \quad \text{and} \quad y_2 = -\frac{R_C L^3}{48EI}$$

Then

$$\Sigma y = \frac{(5W - 8R_C)L^3}{384EI} = 1.25 \text{ in.}$$

Substituting numerical values and solving,

$$R_C = \textbf{49,800 lb}$$
$$R_L = R_R = \tfrac{1}{2}(136,800 - 49,800) = 43,500 \text{ lb}$$
$$M_C = 43,500 \times 18 - 3800 \times 18 \times 9 = \textbf{167,400 ft-lb}$$

Refer to the shear diagram.

$$V_A = 43,500 - 3800x = 0 \qquad \therefore x = 11.45 \text{ ft}$$
$$M_A = M_{\max} = \tfrac{1}{2} \times 43,500 \times 11.45 = 249,000 \text{ ft-lb}$$

$$f = \frac{249,000 \times 12}{151} = \textbf{19,800 psi}$$

7-7   Theorem of Three Moments.   In Fig. 7-8, ab and bc are two adjacent spans of a continuous beam, the supports at a, b, and c being at the same level. Let M_a, M_b, and M_c denote the bending moments at a, b, and c, respectively. If the beam is prismatic and carries the indicated load system, it can be demonstrated that these bending moments and the load system are related by the following equation:

$$\begin{aligned}
M_a L_1 + 2M_b(L_1 + L_2) + M_c L_2 = & -\tfrac{1}{4}w_1 L_1{}^3 - \tfrac{1}{4}w_2 L_2{}^3 \\
& - P_1 L_1{}^2 k_1(1 + k_1)(1 - k_1) \\
& - P_2 L_2{}^2 k_2(1 - k_2)(2 - k_2)
\end{aligned} \tag{7-6}$$

If a span contains several concentrated loads, the corresponding term of this equation must be applied for each load. It is to be noted that k_1 and k_2 are pure numbers.

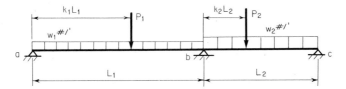

FIG. 7-8.   Adjacent spans of continuous beam.

The foregoing relationship, which is known as the *theorem of three moments*, affords a means of analyzing a continuous beam that is prismatic, rests on simple supports in one horizontal plane, and carries the type of load system indicated. The technique consists of the following steps:

1. Take the first and second spans and substitute in Eq. (7-6). Then take the second and third spans and substitute in this equation; etc. This cyclic procedure generates a system of simultaneous equations containing the bending moments at the interior supports.

2. Solve this system of simultaneous equations.

3. Set up the expression for the bending moment at each interior support and equate this moment to the value obtained in step 2. Solve the resulting system of equations to obtain the vertical reactions at the supports.

If the beam carries a load system that differs from that shown in Fig. 7-8, it is theoretically possible to derive an equation analogous to Eq. (7-6), but in practice the equation may be too cumbersome to be useful.

Example 7-7.   Determine the reactions for the prismatic beam in Fig. 7-9a.

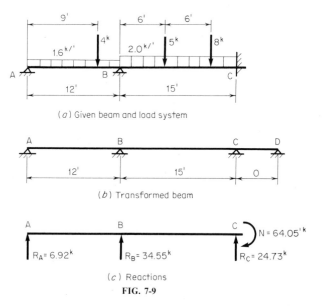

(*a*) Given beam and load system

(*b*) Transformed beam

(*c*) Reactions

FIG. 7-9

Solution. Since the beam is fixed at its right end, it is not directly amenable to analysis by the theorem of three moments. However, the member can readily be transformed to a beam to which Eq. (7-6) is applicable by means of the following construction:

1. Replace the fixed support at *C* with a simple support, as shown in Fig. 7-9*b*. This substitution allows the right end of the beam to rotate.

2. Extend the beam by adding a span *CD* of arbitrary length. The effect of this span is to restore some restriction on the rotation at *C*.

3. Allow the span length *CD* to approach zero. As it does so, the rotation at *C* also approaches zero. In the limit, when *D* becomes coincident with *C*, the rotation at *C* becomes zero, and the beam in Fig. 7-9*b* is equivalent to the given beam in all respects.

The values pertaining to the loads are as follows:

For the 4-kip load: $\qquad\qquad k = \frac{9}{12} = 0.75$
$$k(1 + k)(1 - k) = 0.75 \times 1.75 \times 0.25 = 0.3281$$

For the 5-kip load: $\qquad\qquad k = \frac{6}{15} = 0.40$
$$k(1 + k)(1 - k) = 0.40 \times 1.40 \times 0.60 = 0.3360$$
$$k(1 - k)(2 - k) = 0.40 \times 0.60 \times 1.60 = 0.3840$$

For the 8-kip load: $\qquad\qquad k = \frac{12}{15} = 0.80$
$$k(1 + k)(1 - k) = 0.80 \times 1.80 \times 0.20 = 0.2880$$
$$k(1 - k)(2 - k) = 0.80 \times 0.20 \times 1.20 = 0.1920$$

For the distributed load on *AB*:

$$\tfrac{1}{4}wL^3 = \tfrac{1}{4} \times 1.6 \times 12^3 = 691.2 \text{ ft}^2\text{-kips}$$

For the distributed load on *BC*:

$$\tfrac{1}{4}wL^3 = \tfrac{1}{4} \times 2.0 \times 15^3 = 1687.5 \text{ ft}^2\text{-kips}$$

The known bending moments are

$$M_A = 0 \qquad \text{and} \qquad M_D = 0$$

First cycle. Take spans *AB* and *BC*. Equation (7-6) yields

$$2M_B(12 + 15) + M_C \times 15 = -691.2 - 1687.5 - 4 \times 12^2 \times 0.3281$$
$$-5 \times 15^2 \times 0.3840 - 8 \times 15^2 \times 0.1920$$

or $\qquad\qquad\qquad 54M_B + 15M_C = -3345 \qquad\qquad\qquad\qquad (a)$

Second cycle. Take spans *BC* and *CD*. Equation (7-6) yields

$$M_B \times 15 + 2M_C \times 15 = -1687.5 - 5 \times 15^2 \times 0.3360 - 8 \times 15^2 \times 0.2880$$

or $\qquad\qquad\qquad 15M_B + 30M_C = -2584 \qquad\qquad\qquad\qquad (b)$

Solving Eqs. (*a*) and (*b*), we obtain

$$M_B = -44.15 \text{ ft-kips} \qquad\qquad M_C = -64.05 \text{ ft-kips}$$

The negative value of M_C signifies that the forces to the left of C exert a counterclockwise moment about C. This moment is balanced by the reaction N of the fixed support, and therefore

$$N = 64.05 \text{ ft-kips clockwise}$$

The total distributed loads are

$$W_{AB} = 1.6 \times 12 = 19.2 \text{ kips} \qquad W_{BC} = 2.0 \times 15 = 30.0 \text{ kips}$$

Refer to Fig. 7-9c. The vertical reactions are found by means of the following calculations:

$$M_B = 12R_A - 19.2 \times 6 - 4 \times 3 = -44.15$$
$$\therefore R_A = 6.92 \text{ kips}$$
$$M_C = 27 \times 6.92 + 15R_B - 19.2 \times 21 - 30.0 \times 7.5 - 4 \times 18$$
$$- 5 \times 9 - 8 \times 3 = -64.05$$
$$\therefore R_B = 34.55 \text{ kips}$$
$$\Sigma F_v = 6.92 + 34.55 + R_C - (19.2 + 30.0 + 4 + 5 + 8) = 0$$
$$\therefore R_C = 24.73 \text{ kips}$$

The vertical reactions can be tested by the following calculation:

$$\Sigma M_A = 19.2 \times 6 + 30.0 \times 19.5 + 4 \times 9 + 5 \times 18 + 8 \times 24$$
$$+ 64.05 - 34.55 \times 12 - 24.73 \times 27 = 0 \qquad \text{OK}$$

7-8   Beam Design Based on Deflection.

Since the excessive deformation of a beam can have deleterious effects, strain is often a criterion in selecting a beam to carry a given load system.

Example 7-8.   A steel beam on a simple span of 36 ft will carry a uniformly distributed load, including its estimated weight, of 1600 plf. The allowable bending stress is 24,000 psi, $E = 29 \times 10^6$ psi, and the deflection is limited to 1 in. Select a section, showing all calculations.

Solution.

$$M = \tfrac{1}{8}wL^2 = \tfrac{1}{8} \times 1600 \times 36^2 \times 12 = 3,110,000 \text{ in.-lb}$$

$$S = \frac{M}{f} = \frac{3,110,000}{24,000} = 129.6 \text{ in.}^3$$

The maximum deflection of a simple beam under a uniform load is

$$y = \frac{5wL^4}{384EI} \qquad \therefore I = \frac{5wL^4}{384Ey}$$

Substituting according to the convention stated in Example 7-4, we have

$$I = \frac{5 \times 1600 \times 36^4 \times 1728}{384 \times 29 \times 10^6 \times 1} = 2085 \text{ in.}^4$$

The steel sections are tabulated at the beginning of Sec. 2 of the AISC Manual in descending order with respect to their section moduli. The lightest beam that meets the strength requirement is W 24 × 61, having $S = 130$ in.^{3} However, referring to Sec. 1 of the AISC Manual, it is found that for this section $I = 1540$ in.^{4}, and therefore the section is deficient with respect to stiffness. Use **W 24 × 76**, having the values

$$S = 176 \text{ in.}^3 \qquad \text{and} \qquad I = 2100 \text{ in.}^4$$

8

TORSION

8-1 Stresses in Cylindrical Shaft. The right circular shaft in Fig. 8-1 has a fixed support at its left end and is free or floating at its right end. If a torque (moment) is applied to the member in the plane of the end section, this torque is transmitted through the member, inducing a resisting torque at the support. Consequently, shearing forces are present on each cross section, the resultant of these forces being a couple having a moment equal to the applied torque.

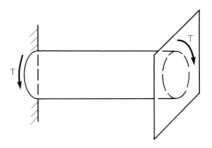

FIG. 8-1. Torque applied to cylindrical shaft.

The torque causes each cross section to rotate about its center, with radial lines remaining straight. It follows that the shearing stress at a given point is directly proportional to the radial distance from that point to the longitudinal axis of the shaft. Thus, the shearing stress is zero at the center and maximum at the circumference. Let

$T =$ applied torque
$J =$ polar moment of inertia of cross-sectional area with respect to longitudinal axis
$s_s =$ shearing stress at given point
$r =$ radial distance from given point to center
$R =$ radius of shaft

Then $\quad\quad s_s = \dfrac{Tr}{J} \quad$ and $\quad s_{s,max} = \dfrac{TR}{J} \quad\quad$ (8-1)

If the shaft is a hollow cylinder having external and internal diameters of D and d, respectively, the equation for maximum shearing stress reduces to

$$s_{s,max} = \frac{16TD}{\pi(D^4 - d^4)} \quad\quad (8\text{-}2)$$

If the shaft is a solid cylinder, $d = 0$ and therefore

$$s_{s,max} = \frac{16T}{\pi D^3} \quad\quad (8\text{-}3)$$

Example 8-1. A torque of 8000 ft-lb is applied to a cylindrical shaft having an outside diameter of 4 in. and inside diameter of 2 in. Calculate the maximum shearing stress.

Solution. Applying Eq. (8-2) and omitting the second subscript for convenience,

$$s_s = \frac{16 \times 8000 \times 12 \times 4}{\pi(256 - 16)} = \textbf{8150 psi}$$

Example 8-2. In applying a wrench to tighten a $\frac{7}{8}$-in. bolt, what torque can a workman apply without exceeding a shearing stress in the bolt of 10,000 psi? The diameter at the root of thread is 0.731 in. Neglect friction.

Solution. By Eq. (8-3),

$$T = 10,000 \times \pi \times 0.731^3/16 = \textbf{767 in.-lb}$$

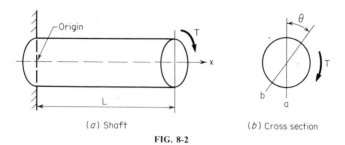

(*a*) Shaft   (*b*) Cross section

FIG. 8-2

8-2   Deformation of Cylindrical Shaft.   With reference to the cylindrical shaft in Fig. 8-2*a*, consider that the *x* axis is placed along the longitudinal axis and the origin of coordinates is placed at the fixed end. As shown in Fig. 8-2*b*, a torque applied at the free end causes each cross section to rotate about its center, so that a diameter that was initially in the position *a* is displaced to the position *b*. The angle of rotation is referred to as the *angle of twist* of the section.

We append the following symbols to the notational system presented in Art. 8-1:

θ = angle of twist of cross section
L = length of shaft
G = modulus of rigidity of material

The angle of twist varies linearly from zero at the fixed end to its maximum value at the free end. If two sections are an infinitesimal distance *dx* apart, the difference between their angles of twist is

$$d\theta = \frac{T}{GJ} \, dx \tag{8-4}$$

Then $\qquad \theta = \dfrac{Tx}{GJ} \qquad$ and $\qquad \theta_{max} = \dfrac{TL}{GJ} \qquad$ (8-5)

For a hollow cylinder, the equation for maximum angle of twist becomes

$$\theta_{max} = \frac{32TL}{G\pi(D^4 - d^4)} \tag{8-6}$$

For a solid cylinder,

$$\theta_{max} = \frac{32TL}{G\pi D^4} \tag{8-7}$$

Example 8-3. A torque of 9000 ft-lb is applied at the free end of a 14-ft cylindrical shaft having an external diameter of 5 in. and internal diameter of 3 in. Compute the angle of twist at the free end, using $G = 6 \times 10^6$ psi. Express the answer in degrees.

Solution. By Eq. (8-6),

$$\theta = \frac{32 \times 9000 \times 12 \times 14 \times 12}{6 \times 10^6 \pi (625 - 81)} = 0.0566 \text{ radian}$$

$$1 \text{ radian} = 57.30°$$
$$\therefore \ \theta = 0.0566 \times 57.30 = \textbf{3.24}°$$

Consider that a hollow shaft is to be designed to transmit a given torque, applying both strength and rigidity as the criteria. The dimensions D and d may be found in this manner: If Eq. (8-2) is divided by Eq. (8-6) and the equation thus obtained is solved for D, the result is

$$D = \frac{2Ls_{s,\max}}{G\theta_{\max}} \tag{8-8}$$

The computed value of D can then be substituted in Eq. (8-2) or (8-6) to obtain d.

Example 8-4. A hollow shaft of 8-ft length is to transmit a torque of 22,000 ft-lb. The shearing stress is restricted to 15,000 psi and the angle of twist is restricted to 3°. Using $G = 12 \times 10^6$ psi, establish the dimensions of the shaft.

Solution

$$L = 96 \text{ in.} \qquad T = 22,000 \times 12 = 264,000 \text{ in.-lb}$$

$$\theta = \frac{3}{57.30} = 0.0524 \text{ radian}$$

By Eq. (8-8),

$$D = \frac{2 \times 96 \times 15,000}{12 \times 10^6 \times 0.0524} = \textbf{4.58 in.}$$

Rearranging Eq. (8-2),

$$d^4 = D^4 - \frac{16TD}{\pi s_s} = 4.58^4 - \frac{16 \times 264,000 \times 4.58}{\pi \times 15,000} = 29.40 \text{ in.}^4$$

$$d = \textbf{2.33 in.}$$

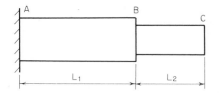

FIG. 8-3. Compound shaft.

The shaft in Fig. 8-3 consists of two distinct parts, AB and BC, connected to one another at their juncture B in such manner that one part cannot rotate relative to the other. This is referred to as a *compound* shaft. Consider that a torque T is applied at the free end C. Let the subscripts 1 and 2 refer to the parts AB and BC, respectively. The angle of twist at C is found by integrating Eq. (8-4) across the entire shaft, giving

$$\theta = T\left(\frac{L_1}{G_1 J_1} + \frac{L_2}{G_2 J_2}\right) \tag{8-9}$$

If both parts are solid, this equation becomes

$$\theta = \frac{32T}{\pi}\left(\frac{L_1}{G_1 D_1{}^4} + \frac{L_2}{G_2 D_2{}^4}\right) \tag{8-10}$$

8-3 Statically Indeterminate Shafts. An advantage accrues in recasting Eq. (8-5) in the following form:

$$\theta_{max} = \frac{T}{GJ/L}$$

Since the angle of twist is inversely proportional to GJ/L, we may regard this quantity as the resistance that the shaft offers to deformation. Therefore, GJ/L represents the *torsional stiffness* of the member, denoted by K_t.

The shaft AB in Fig. 8-4 is connected to a fixed support at each end. Consider that a torque T is applied to the member at some intermediate section Q. This torque divides into two elements: torques T_A and T_B transmitted from Q to A and from Q to B, respectively. The elements induce the resisting torques shown in the drawing.

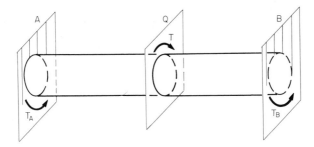

FIG. 8-4. Shaft fixed at both ends.

The shaft *AB* may be visualized as consisting of two distinct parts, *AQ* and *BQ*, which are fixed at *A* and *B*, respectively, free at *Q*, but constrained to undergo an identical rotation at *Q*. By setting up the expression for the angle of twist at *Q* for each part of the shaft and then equating these expressions, it is found that the torque divides into elements that are directly proportional to the stiffnesses of the respective parts. Thus,

$$\frac{T_A}{T_B} = \frac{K_{t,AQ}}{K_{t,BQ}} \qquad \text{and} \qquad T_A + T_B = T \qquad (8\text{-}11)$$

Example 8-5. The compound solid shaft in Fig. 8-5 is fixed at both ends, and it has the dimensions and properties shown in the drawing. A torque of 100,000 in.-lb is applied at *C*. Find the maximum shearing stress in the member.

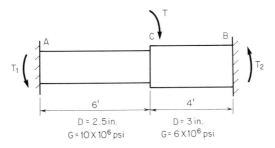

FIG. 8-5

Solution. Let the subscripts 1 and 2 refer to the parts AC and BC, respectively. By Eq. (8-7), the angle of twist at C is

$$\theta_C = \frac{32T_1L_1}{G_1\pi D_1{}^4} = \frac{32T_2L_2}{G_2\pi D_2{}^4}$$

$$\therefore \quad \frac{T_1}{T_2} = \frac{G_1}{G_2}\left(\frac{D_1}{D_2}\right)^4 \frac{L_2}{L_1} = \frac{10}{6}\left(\frac{2.5}{3}\right)^4 \frac{4}{6} = 0.536$$

and
$$T_1 = 0.536T_2$$

Then
$$T_1 + T_2 = 1.536T_2 = 100,000 \text{ in.-lb}$$

$$T_2 = 65,100 \text{ in.-lb} \qquad T_1 = 100,000 - 65,100 = 34,900 \text{ in.-lb}$$

Apply Eq. (8-3).

In AC:
$$s_s = \frac{16 \times 34,900}{\pi \times 2.5^3} = 11,400 \text{ psi}$$

In BC:
$$s_s = \frac{16 \times 65,100}{\pi \times 3^3} = \textbf{12,300 psi} = s_{s,\text{max}}$$

8-4   Torsion on Rivet Group.   Plate AB in Fig. 8-6a is riveted to its support and is subjected to a torque caused by the applied couple. This torque is transmitted from the plate to the rivets and then to the support. The rivets therefore sustain shearing stresses.

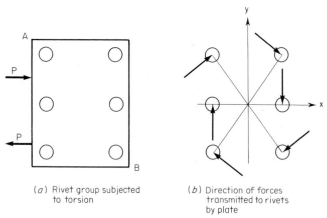

(*a*) Rivet group subjected   (*b*) Direction of forces
to torsion   transmitted to rivets
                                  by plate

FIG. 8-6

If a plane is passed through the rivets parallel to the plate, the cross-sectional area of the rivets may be considered to form an aggregate (but discontinuous) area. Although this aggregate rivet area is not truly cylindrical, no significant error will be injected if we apply Eq. (8-1) in calculating the shearing stresses in the rivets. For convenience, assume that the resultant force on a rivet acts at the center of the rivet.

The simplifications we have introduced lead to the following conclusions:

1. The force on a rivet is directly proportional to its distance from the centroid of the aggregate rivet area.

2. The force on a rivet has a direction normal to the radius from the centroid of the aggregate area to the center of rivet, as indicated by the vectors in Fig. 8-6b.

Establish the coordinate axes shown in Fig. 8-6b, placing the origin at the centroid of the aggregate rivet area. Assume that all rivets in the group are of the same size, and take the cross-sectional area of a rivet as a unit of area. Let

T = torque transmitted through connection

f = force on given rivet caused by this torque

r = radial distance from origin to center of given rivet

The polar moment of inertia of the aggregate area is

$$J = \Sigma r^2 = \Sigma(x^2 + y^2)$$

where the summation includes all rivets in the group.

Equation (8-1) assumes the following form in the present instance:

$$f = \frac{Tr}{\Sigma(x^2 + y^2)} \qquad (8\text{-}12)$$

Consider f to be resolved into its x and y components, and let x and y denote the coordinates of the given rivet. Consider a clockwise torque as positive. The algebraic values of the components of f are as follows:

$$f_x = \frac{Ty}{\Sigma(x^2 + y^2)} \qquad \text{and} \qquad f_y = -\frac{Tx}{\Sigma(x^2 + y^2)} \qquad (8\text{-}13)$$

Where the torque on a rivet group results from eccentricity of loading, the AISC Manual recommends use of a reduced eccentricity in the calculations in recognition of certain secondary effects. However, we shall apply the true eccentricity in the examples that follow. A concentric load applied to a rivet group is assumed to induce a uniform shearing stress in the group.

Example 8-6.   The bracket in Fig. 8-7 carries a load P of 10,000 lb. Compute the maximum force on a rivet, and identify the rivet that receives this force.

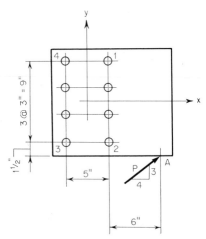

FIG. 8-7. Eccentric load on riveted connection.

Solution.   It is understood that the rivets are all of the same size. Therefore, by symmetry, the centroid of the aggregate rivet area lies midway between the two vertical rows and midway between the two inner horizontal rows. Then

$$\Sigma(x^2 + y^2) = 8 \times 2.5^2 + 4 \times 1.5^2 + 4 \times 4.5^2 = 140 \text{ in.}^2$$

Resolve P into its x and y components, giving

$$P_x = 8000 \text{ lb} \qquad P_y = 6000 \text{ lb}$$

Consider these components acting at A. Their eccentricities from the x and y axes are

$$e_x = 1.5 + 3 + 1.5 = 6 \text{ in.} \qquad e_y = 2.5 + 6 = 8.5 \text{ in.}$$

The load P can therefore be transformed to an equivalent system consisting of concentric horizontal and vertical loads of 8000 lb and 6000 lb, respectively, and a torque having the value

$$T = -8000 \times 6 - 6000 \times 8.5 = -99,000 \text{ ft-lb}$$

Let f = force on rivet due to torque
    f' = force on rivet due to concentric loading
    F = resultant force on rivet

Label the four outermost rivets in the manner shown. It will be found that rivet 2 is the most heavily loaded in the group because it is the only rivet among those labeled for which corresponding components of f and f' have the same algebraic sign. (It should also be noted that rivet 2 lies closer to the action line of P than do the other labeled rivets.) By Eq. (8-13),

$$f_x = \frac{(-99,000)(-4.5)}{140} = 3180 \text{ lb}$$

$$f_y = -\frac{(-99,000)2.5}{140} = 1770 \text{ lb}$$

The x and y components of f' are both positive.

$$f'_x = \frac{8000}{8} = 1000 \text{ lb} \qquad f'_y = \frac{6000}{8} = 750 \text{ lb}$$

$$F_x = 3180 + 1000 = 4180 \text{ lb} \qquad F_y = 1770 + 750 = 2520 \text{ lb}$$
$$F = \sqrt{4180^2 + 2520^2} = \textbf{4880 lb}$$

Part 4 of the AISC Manual presents tables to expedite the design of a rivet group under torsion. Another table presents values of the net section modulus of a plate that has vertical rows of rivets with a 3-in. spacing.

8-5 Beam-Shafts. Many structural members function in the dual role of beams and shafts. The maximum stresses in these members are found by applying the equations in Art. 3-3.

Example 8-7. The horizontal member in Fig. 8-8 has an external diameter of 4 in. and an internal diameter of 3 in. It carries a vertical load of 1600 lb at the location shown. Neglecting the weight of the member, compute the principal stresses and maximum shearing stress at a, which lies at the top of shaft at the wall.

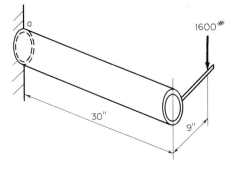

FIG. 8-8. Beam-shaft.

Solution. The given load is transformed to an equivalent system consisting of a concentric vertical load of 1600 lb applied at the free end and a torque of

$$T = 1600 \times 9 = 14{,}400 \text{ in.-lb}$$

Beam action. The section modulus is

$$S = \frac{\pi(D^4 - d^4)}{32D} = \frac{\pi(4^4 - 3^4)}{32 \times 4} = 4.30 \text{ in.}^3$$

At the support, $M = 1600 \times 30 = 48{,}000 \text{ in.-lb}$

Then $f = \dfrac{M}{S} = \dfrac{48{,}000}{4.30} = 11{,}160 \text{ psi}$

$$v = 0$$

Shaft action. Rearranging Eq. (8-2) gives

$$s_s = \frac{T}{2S} = \frac{14{,}400}{2 \times 4.30} = 1670 \text{ psi}$$

Figure 8-9 is a free-body diagram of an infinitesimally small portion of the shaft at *a*. Applying the notational system of Fig. 3-4*b*, we have

$$f_1 = 0 \qquad f_2 = 11{,}160 \text{ psi} \qquad s_{s1} = 1670 \text{ psi}$$

$$\frac{f_1 + f_2}{2} = 5580 \text{ psi} \qquad \frac{f_1 - f_2}{2} = -5580 \text{ psi}$$

By Eqs. (3-3),

$$s_{s,\max} = R = \sqrt{5580^2 + 1670^2} = \textbf{5820 psi}$$
$$f_{\max} = 5580 + 5820 = \textbf{11,400 psi (tension)}$$
$$f_{\min} = 5580 - 5820 = \textbf{-240 psi (compression)}$$

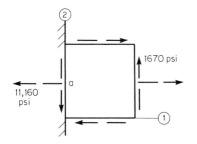

FIG. 8-9. Stresses on longitudinal and transverse planes at *a*.

PROBLEMS

8-1. A hollow cylindrical shaft has an internal diameter of 4 in. and a length of 88 in. The modulus of rigidity is 9,000,000 psi. A torque of 25,000 ft-lb applied at the free end induces a shearing stress of 8600 psi. Determine the external diameter of the shaft and the angle of twist at the free end. ANS. $D = 6.04$ in.; $\theta = 1.61°$

8-2. The solid compound shaft in Fig. 8-3 has the following data: for AB, $L = 5$ ft and $D = 4$ in.; for BC, $L = 3.5$ ft and $D = 3$ in. For the entire shaft, $G = 12 \times 10^6$ psi. What torque may be applied at C if the shearing stress is limited to 15,000 psi and the angle of twist is limited to $2.5°$? ANS. 68,300 in.-lb

8-3. With reference to the compound shaft in Fig. 8-5, what torque must be applied at C to induce an angle of twist of $2°$ at a section 15 in. to the right of C?
ANS. 77,500 in.-lb

8-4. With reference to the riveted connection in Fig. 8-10, the maximum force on a rivet caused by the load P is 5800 lb. Identify this rivet, and determine the magnitude of P. ANS. $P = 35,580$ lb

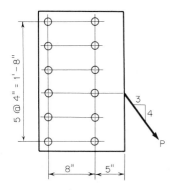

FIG. 8-10

9

STEEL MEMBERS IN TENSION AND COMPRESSION

9-1   Design of Tension Members.   If the plate in Fig. 9-1*a* is connected to its support with two rivets located in the transverse section *AB*, its capacity as a tension member is the product of the allowable tensile stress and the net area of the plate at *AB*. Now consider that the lower rivet is displaced a small distance *s* to the right, as shown in Fig. 9-1*b*. The capacity of the plate has manifestly been enhanced, and it is necessary to formulate a rule for gaging the capacity when the rivets are staggered.

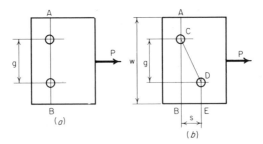

FIG. 9-1. Net area of tension member.

Section 1.14.3 of the AISC Specification prescribes that we compute the net width at section $ACDE$ by deducting the diameters of two holes from the gross width and then adding the quantity $s^2/4g$. For example, assume the following values with reference to Fig. 9-1b:

$$w = 6 \text{ in.} \qquad g = 3 \text{ in.} \qquad s = 2 \text{ in.}$$
$$\text{Rivet diameter} = \tfrac{3}{4} \text{ in.}$$

There are two sections of potential rupture: AB and the irregular section $ACDE$. Taking the diameter of the holes as $\tfrac{7}{8}$ in., we obtain the following net widths:

$$w_{AB} = 6 - \tfrac{7}{8} = 5.13 \text{ in.}$$

$$w_{ACDE} = 6 - 2 \times \frac{7}{8} + \frac{2^2}{4 \times 3} = 4.58 \text{ in.}$$

Thus the plate is more likely to rupture along section $ACDE$ than AB.

If an irregular section of potential rupture crosses several rivets, we deduct from the gross width the sum of the diameters of all holes in the section and add the aggregate of the $s^2/4g$ values between successive rivets.

Example 9-1. The 8- by $\tfrac{5}{16}$-in. plate in Fig. 9-2 sustains an axial tensile force of 30,000 lb and is riveted to its support with three $\tfrac{7}{8}$-in.-diameter rivets in the manner shown. If the allowable tensile stress is 22,000 psi, is the section adequate?

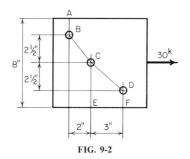

FIG. 9-2

Solution.   There are two sections of potential rupture: *ABCE* and *ABCDF*.

$$w_{ABCE} = 8 - 2 \times 1 + \frac{2^2}{4 \times 2.5} = 6.4 \text{ in.}$$

$$w_{ABCDF} = 8 - 3 \times 1 + \frac{2^2}{4 \times 2.5} + \frac{3^2}{4 \times 2.5} = 6.3 \text{ in.}$$

$$f = \frac{30,000}{6.3 \times 0.313} = 15,200 \text{ psi}$$

∴ Section is adequate.

Example 9-2.   The bottom chord of a roof truss resists a tensile force of 139,300 lb. The member will be spliced with $\frac{3}{4}$-in.-diameter rivets in the manner shown in Fig. 9-3a. Design a double-angle member, using an allowable stress of 22,000 psi.

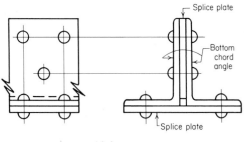

(a)

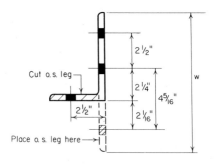

(b)

FIG. 9-3. Net area of angle in tension.

Solution.   Although there will be three gage lines in each angle, the loss of area can be reduced by staggering the rivets. We shall therefore assume a deduction of $2\frac{1}{2}$ rivet holes. It is convenient to visualize each angle in its "developed" form, i.e., as it would appear if the two legs were severed and made coplanar, as shown in Fig. 9-3b. Section 1.14.4 of the AISC Specification is applicable to this problem. The angle size can be arrived at by a trial-and-error method or by the direct method that follows. Let w and t denote, respectively, the gross width and thickness of each angle. Then

$$\text{Net area} = 2t(w - 2.5 \times 0.875) = 139{,}300/22{,}000 = 6.33 \text{ sq in.}$$

$$\therefore\; w = \frac{3.17}{t} + 2.19$$

$$\begin{array}{ll} \text{If } t = \tfrac{1}{2} \text{ in.}, & w = 8.53 \text{ in.} \\ \text{If } t = \tfrac{7}{16} \text{ in.}, & w = 9.44 \text{ in.} \end{array}$$

The first condition is satisfied by $6 \times 3\frac{1}{2} \times \frac{1}{2}$ in. angles ($w = 9$ in.), and the second by $6 \times 4 \times \frac{7}{16}$ in. angles ($w = 9.56$ in.). The latter size is more economical.

$\therefore$ Use two angles **$6 \times 4 \times \frac{7}{16}$ in.**

Although the problem does not require that this be done, we shall compute the minimum allowable value of the rivet stagger s.

$$\begin{array}{ll} \text{Gross area of two angles} \dots\dots\dots\dots\dots & 8.36 \text{ sq in.} \\ \text{Net area required} \dots\dots\dots\dots\dots\dots & 6.33 \\ \text{Allowable reduction of area} \dots\dots\dots\dots & 2.03 \text{ sq in.} \end{array}$$

$$\text{Allowable reduction of width} = \frac{2.03}{\frac{7}{16}} = 4.64 \text{ in.}$$

The standard gages for angles are recorded in Part 4 of the AISC Manual. In Fig. 9-3b,

$$3 \times \frac{7}{8} - \frac{s^2}{4 \times 2.5} - \frac{s^2}{4 \times 4.31} = \frac{4.64}{2}$$

Solving, $$s_{\min} = 1.4 \text{ in.}$$

9-2   Columns with Axial Loads.

A *column* is a compression member having a length that is large in relation to its lateral dimensions. As the axial forces applied to the column are increased, the member eventually buckles. The *effective length* of a column is the distance between adjacent points of inflection in the buckled member or in the imaginary extension of the buckled member. The column length is denoted by L and the effective length by KL. Recommended values of K are presented in the Commentary on the AISC Specification.

The capacity of a column is a function of its effective length and the properties of its cross section. It therefore becomes necessary to extend our analysis of the properties of an area.

Consider that the moment of inertia I of an area is evaluated with respect to a group of concurrent axes (i.e., axes passing through a common point). There is a distinct value of I associated with each axis. The *major* axis is the one for which I is maximum; the *minor* axis is the one for which I is minimum. The major and minor axes are referred to collectively as the *principal* axes.

The following theorems may be readily proved:

a. The principal axes through a given point are mutually perpendicular.

b. An axis of symmetry is a principal axis.

c. If several areas all have the same radius of gyration with respect to a given axis, the radius of gyration of their composite area equals that of the individual areas.

In the absence of any restraint, a column tends to buckle about the minor centroidal axis of its cross section. Consequently, the capacity of a column is a function of its minimum radius of gyration. The first step in the investigation of a column therefore consists of identifying the minor centroidal axis and computing the corresponding radius of gyration r. The ratio KL/r is called the *effective slenderness ratio.* Section 1.5.1.3 of the AISC Specification provides two equations for the allowable stress in a column; the one to be applied depends on the relationship of the effective slenderness ratio to the quantity C_c. Rather than substitute in these equations, we shall determine the allowable stress by referring to the tables in Appendix A of the AISC Specification. With reference to the rolled section in Fig. 9-4, XX is the major axis and YY is the minor axis.

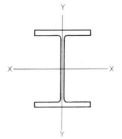

FIG. 9-4. Principal axes.

In the ensuing material, A denotes the cross-sectional area and P the column capacity. In the absence of any statement to the contrary, it is to be understood that all loads are axial, all members are made of A36 steel, and the effective length coincides with the actual length.

Example 9-3. Select an economical rolled-steel section for a 20-ft column to carry a load of 620 kips.

Solution. Part 3 of the AISC Manual presents tables of allowable concentric loads for columns. Referring to these tables, we find that a W 14 × 127 column with an effective length of 20 ft has a capacity of 636 kips. We shall verify the selection. The area and minimum radius of gyration of the section are

$$A = 37.3 \text{ sq in.} \qquad r = 3.76 \text{ in.}$$

$$\frac{KL}{r} = \frac{20 \times 12}{3.76} = 63.8$$

The allowable stress as given in Table 1-36 in Part 5 of the AISC Manual is 17.06 ksi. Then

$$P = Af = 37.3 \times 17.06 = 636 \text{ kips}$$

∴ Use a **W 14 × 127**.

Example 9-4. A riveted H column, 22 ft long, is composed of four angles 6 × 4 × $\frac{1}{2}$ in. and a web plate 10 by $\frac{1}{2}$ in. The angles are set $10\frac{1}{2}$ in. back to back with the long legs outstanding. Compute the capacity of the member.

Solution. In accordance with Sec. 1.14.2 of the AISC Specification, the gross area of a compression member is applied in computing its capacity. Refer to Fig. 9-5. It is evident that YY is the minor axis, but this conclusion must be verified.

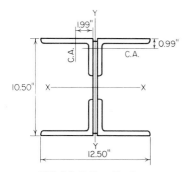

FIG. 9-5. Built-up H column.

Calculation of I_x

$$\text{One plate,} \quad I_o' = \tfrac{1}{12} \times \tfrac{1}{2} \times 10^3 = \quad 41.7 \text{ in.}^4$$
$$\text{Four angles,} \quad I_o' = 4 \times 6.3 = \quad 25.2$$
$$Ak'^2 = 4 \times 4.75(5.25 - 0.99)^2 = \underline{344.8}$$
$$I_x = 411.7 \text{ in.}^4$$

Calculation of I_y

$$\text{Four angles,} \quad I_o' = 4 \times 17.4 = \quad 69.6 \text{ in.}^4$$
$$Ak'^2 = 4 \times 4.75(1.99 + 0.25)^2 = \underline{95.3}$$
$$I_y = 164.9 \text{ in.}^4$$

∴ Set $I = 164.9 \text{ in.}^4$

$$A = \tfrac{1}{2} \times 10 + 4 \times 4.75 = 24.0 \text{ sq in.}$$

$$r = \sqrt{\frac{164.9}{24.0}} = 2.62 \text{ in.} \qquad \frac{KL}{r} = \frac{22 \times 12}{2.62} = 100.8$$

By Table 1-36 of the Manual, the allowable stress is 12.88 ksi.

$$P = 24.0 \times 12.88 = \mathbf{309 \ kips}$$

Example 9-5. A compression member consists of two C 15 × 40 channels laced together and spaced 10 in. back to back with flanges outstanding. The member is 15 ft long. What load can the member carry?

Solution. Refer to Fig. 9-6. We must first identify the critical axis. The radius of gyration of the member about XX equals that of the individual

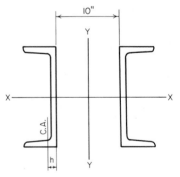

FIG. 9-6. Built-up column.

channel, and the radius of gyration about YY is obtained by applying Eq. (2-11). For a single channel:

$$A = 11.8 \text{ sq in.} \qquad h = 0.78 \text{ in.}$$
$$r_1 = 5.44 \text{ in.} \qquad r_2 = 0.89 \text{ in.}$$

For the built-up member:

$$r_x = 5.44 \text{ in.} \qquad r_y = \sqrt{r_2{}^2 + 5.78^2} > 5.78 \text{ in.}$$
$$\therefore r = 5.44 \text{ in.} \qquad \frac{KL}{r} = \frac{15 \times 12}{5.44} = 33.1$$

By Table 1-36, $\qquad f = 19.72 \text{ ksi}$
Then $\qquad P = 2 \times 11.8 \times 19.72 = \textbf{465 kips}$

The capacity of a column can be augmented by introducing ties at intermediate points to reduce the effective length KL. In such a situation, the critical section for buckling may not be readily apparent. Let $K_x L$ and $K_y L$ represent the effective lengths with respect to the major and minor axes, respectively. If a column has equal strength in all directions,

$$\frac{K_x L}{r_x} = \frac{K_y L}{r_y} \qquad \text{or} \qquad K_x L = K_y L \frac{r_x}{r_y} \qquad (9\text{-}1)$$

The tables of allowable column loads in the AISC Manual are constructed by assuming that the column tends to buckle about the minor centroidal axis. However, if the length $K_x L$ in a given case exceeds the value given by Eq. (9-1), buckling will occur about the major axis. To expedite the selection of a member where the column has unequal values of $K_x L$ and $K_y L$, the allowable-column-load tables in the AISC Manual provide the values of r_x/r_y for the rolled sections.

Example 9-6. A column that is 30 ft long is to carry a load of 200 kips. The member will be braced about both principal axes at top and bottom and in addition will be braced about its minor axis at mid-height. Select a section having a nominal depth of 8 in. with $f_y = 50$ ksi.

Solution. The effective length with respect to the minor axis may be taken as 15 ft. Then

$$K_x L = 30 \text{ ft} \qquad K_y L = 15 \text{ ft}$$

Assume tentatively that the effective slenderness ratio is greater about the minor axis. Referring to the allowable-column-load tables in the AISC Manual, we find by linear interpolation that a W 8 × 40 column has a capacity of 200 kips when $K_yL = 15.3$ ft and that for this section $r_x/r_y = 1.73$. For uniform strength, the following limit applies:

$$K_xL = 1.73 \times 15.3 = 26.5 < 30 \text{ ft}$$

The section is therefore inadequate. Try a W 8 × 48, which has a capacity of 200 kips when $K_yL = 17.7$ ft. For uniform strength,

$$K_xL = 1.74 \times 17.7 = 30.8 > 30 \text{ ft}$$

The W 8 × 48 therefore appears to be satisfactory, but we shall verify the design. For this section,

$$A = 14.1 \text{ sq in.} \qquad r_x = 3.61 \text{ in.} \qquad r_y = 2.08 \text{ in.}$$

$$\frac{K_xL}{r_x} = \frac{30 \times 12}{3.61} = 100 \qquad \frac{K_yL}{r_y} = \frac{15 \times 12}{2.08} = 87$$

Entering Table 1-50 in Appendix A of the AISC Specification with an effective slenderness ratio of 100, we find that the allowable stress is 14.71 ksi. Then

$$P = 14.1 \times 14.71 = 207 \text{ kips}$$

∴ Use **W 8 × 48.**

9-3   Investigation of Beam-Column.   A member that is subjected concurrently to axial loading and bending is referred to as a *beam-column.* The simultaneous set of values of axial stress and bending stress must satisfy the relationships set forth in Sec. 1.6 of the AISC Specification. Let

f_a = axial stress
f_b = bending stress
F_a = allowable axial stress if member resisted axial loads exclusively
F_b = allowable bending stress if member resisted flexure exclusively
F'_e = stress defined in Sec. 1.6 of Specification
C_m = coefficient defined in Sec. 1.6 of Specification

Example 9-7.   A W 12 × 53 column of 20-ft length and A36 steel is to resist the axial forces and end moments shown in Fig. 9-7. The member will be restrained against sidesway at top and bottom. Is the section adequate?

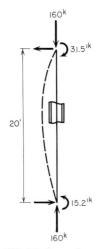

FIG. 9-7. Beam-column.

Solution.

$$A = 15.6 \text{ sq in.} \qquad S_x = 70.7 \text{ in.}^3$$
$$r_x = 5.23 \text{ in.} \qquad r_y = 2.48 \text{ in.}$$

From the allowable-column-load tables,

$$L_c = 10.6 \text{ ft} \qquad\qquad L_u = 22.2 \text{ ft}$$

$$f_a = \frac{160}{15.6} = 10.26 \text{ ksi} \qquad f_b = \frac{31.5 \times 12}{70.7} = 5.35 \text{ ksi}$$

$$\frac{KL}{r} = \frac{240}{2.48} = 96.8 \qquad \therefore F_a = 13.38 \text{ ksi}$$

$$L_u > KL > L_c \qquad\qquad \therefore F_b = 22 \text{ ksi}$$

$$\frac{f_a}{F_a} = \frac{10.26}{13.38} = 0.767 > 0.15$$

The section must therefore satisfy the following requirements:

$$\frac{f_a}{F_a} + \frac{C_m}{1 - f_a/F'_e} \frac{f_b}{F_b} \leqslant 1 \qquad\qquad (9\text{-}2)$$

$$\frac{f_a}{0.6f_y} + \frac{f_b}{F_b} \leqslant 1 \qquad\qquad (9\text{-}3)$$

Since the member is bent in single curvature, the ratio M_1/M_2 is taken as negative, and

$$C_m = 0.6 + 0.4 \times \frac{15.2}{31.5} = 0.793$$

$$F'_e = \frac{149{,}000 \times 5.23^2}{240^2} = 70.76 \text{ ksi}$$

$$\frac{f_a}{F'_e} = \frac{10.26}{70.76} = 0.145$$

Equation (9-2) yields

$$0.767 + \frac{0.793}{0.855} \times \frac{5.35}{22} = 0.993 \qquad \text{OK}$$

Equation (9-3) yields

$$\frac{10.26}{22} + \frac{5.35}{22} = 0.709 \qquad \text{OK}$$

The section is therefore adequate.

10

TRUSSES

10-1   Stability of the Triangle.   The quadrilateral frame *abcd* in Fig. 10-1a is constructed with flexible joints, thereby allowing the members at each joint to rotate relative to one another. Assume that the frame is weightless, and consider that it is supported at *a* and *b* while a load is applied at *c*. This load causes the frame to collapse. Geometrically, failure of the frame is explained by the fact that there is an infinite number of quadrilaterals having sides of the given lengths. Therefore, since the quadrilateral is capable of assuming other shapes, it permits displacement of joint *c* under the load.

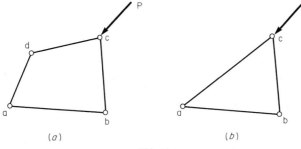

FIG. 10-1

The triangular frame *abc* in Fig. 10-1*b* is also constructed with flexible joints. Consider that the frame is supported at *a* and *b* while a load is applied at *c*. Since the triangle is incapable of assuming other shapes, it resists displacement of joint *c*, and therefore it supports the load.

The foregoing analysis leads to the following conclusion: The triangle is the sole type of hinged polygon that possesses structural stability. Therefore, triangles can serve as building blocks in devising hinged structures.

10-2   Definition of a Truss.   A *truss* is a hinged frame composed of triangles. Thus, the frame in Fig. 10-2*a* is a truss, but the frame in Fig. 10-2*b* is not because *abde* is a quadrilateral. Figure 10-3 illustrates certain definitions pertaining to a standard type of roof truss.

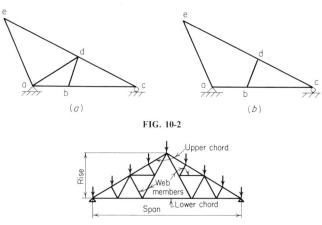

FIG. 10-2

FIG. 10-3. Fink truss.

In theory, a truss is constructed and loaded in the following manner:

1. All joints are formed with frictionless pins.

2. A truss member begins at one joint and terminates at the next joint. Thus, no member is continuous across a joint.

3. The longitudinal axis of each member at a joint passes through the center of the pin.

4. The truss is weightless, and all forces on the truss act at the joint centers.

Figure 10-4 is a free-body diagram of a truss member. On the basis of statements 3 and 4, it is seen that the member is subjected to two forces, each applied at the centroid of the end section. Since the member is in equilibrium, these two forces are numerically equal but oppositely sensed, and they are collinear. Therefore, in theory every truss member is subjected purely to axial loading and not to bending.

FIG. 10-4. Forces on truss member.

Realistically, trusses are constructed and loaded in ways that often differ substantially from those we have specified. Moreover, the weight of truss may not be negligible. Consequently, every truss member sustains stresses beyond those corresponding to the ideal conditions; these are termed *secondary* stresses.

10-3   Truss Analysis.   After the force system acting on a truss has been fully established, it becomes necessary to evaluate the internal forces. The axial force in a given member can be found by cutting the truss through that member and then applying the equations of equilibrium to either part of the truss. In the *method of sections*, the truss is cut along a plane; in the *method of joints*, the truss is cut along a circular arc centered at the joint. However, in a given problem, we are free to amalgamate the two methods. Moreover, the truss can also be cut along an irregular or zig-zag surface if that should prove convenient. As stated in Art. 1-5, a concurrent force system has only two independent equations of equilibrium.

To maintain consistency in our calculations, we shall assume that all internal forces are tensile. Therefore, a positive result signifies that the member is in tension, and a negative result signifies that the member is in compression. We shall continue to apply the sign convention for forces and moments given in Art. 1-3.

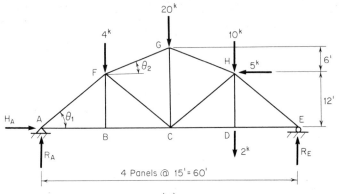

(*a*)

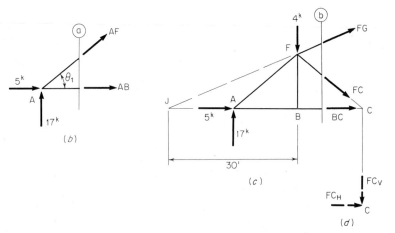

(*b*)

(*c*)

(*d*)

FIG. 10-5

Example 10-1. With reference to the truss in Fig. 10-5*a*, find the force in members *AB* and *FC*, using two completely independent methods in each case.

Solution

$$\Sigma F_H = H_A - 5 = 0 \qquad \therefore H_A = 5 \text{ kips}$$
$$\Sigma M_E = 60R_A - 4 \times 45 - 20 \times 30 - (10 + 2)15 - 5 \times 12 = 0$$
$$\therefore R_A = 17 \text{ kips}$$
$$\tan \theta_1 = 12/15 = 0.800 \qquad \tan \theta_2 = 6/15 = 0.400$$
$$\sin \theta_1 = 0.625 \qquad \cos \theta_1 = 0.781$$

The force in a member will be given the same designation as the member itself, and the subscripts H and V will be used to designate, respectively, the horizontal and vertical components of the force. The character of the force will be specified by using the letters T and C to denote tension and compression, respectively.

To find the force in AB, cut the truss along a plane a through the first panel. Figure 10-5b is the free-body diagram of that part of the truss lying to the left of plane a. If AB and AF are in tension, the forces acting on the parts of the members to the left of the plane are sensed as shown. The force in AB is found by applying the equations of equilibrium to the force system shown in Fig. 10-5b. In the first method, we shall apply the equations

$$\Sigma F_V = 17 + AF_V = 0 \tag{a}$$

and

$$\Sigma F_H = 5 + AF_H + AB = 0 \tag{b}$$

From Eq. (a),

$$AF_V = -17 \text{ kips}$$

Then

$$AF_H = \frac{AF_V}{\tan \theta_1} = \frac{-17}{0.800} = -21.25 \text{ kips}$$

Substituting in Eq. (b), $AB = \mathbf{16.25 \text{ kips T}}$

In the second method, we shall take moments with respect to F, giving

$$\Sigma M_F = 17 \times 15 - 5 \times 12 - 12AB = 0$$
$$\therefore \ AB = 16.25 \text{ kips T}$$

The second method is clearly the simpler of the two.

To find the force in FC, cut the truss along a plane b through the second panel. Figure 10-5c is the free-body diagram of that part of the truss lying to the left of plane b. Three unknown forces are present: FG, FC, and BC. In the first method, we shall secure an equation containing FC alone. Prolong members FG and BC to their intersection at J, and take moments with respect to this point. To obviate the need to calculate the lever arm of FC with respect to J, resolve FC into its horizontal and vertical components. Perform this resolution at C, as shown in Fig. 10-5d. The location of J is as follows:

$$JB = \frac{BF}{\tan \theta_2} = \frac{12}{0.4} = 30 \text{ ft}$$

$$\Sigma M_J = -17 \times 15 + 4 \times 30 + 45FC_V = 0$$
$$\therefore \ FC_V = 3 \text{ kips}$$

$$FC = \frac{FC_V}{\sin \theta_1} = \frac{3}{0.625} = \mathbf{4.80 \text{ kips T}}$$

In the second method, we shall first evaluate FG_H and BC. Resolve FG into its horizontal and vertical components, performing the resolution at G. Then

$$\Sigma M_C = 17 \times 30 - 4 \times 15 + 18FG_H = 0 \qquad \therefore FG_H = -25 \text{ kips}$$
$$\Sigma M_F = 17 \times 15 - 5 \times 12 - 12BC = 0 \qquad \therefore BC = 16.25 \text{ kips}$$
$$\Sigma F_H = 5 + 16.25 - 25 + FC_H = 0 \qquad \therefore FC_H = 3.75 \text{ kips}$$

$$FC = \frac{3.75}{0.781} = 4.80 \text{ kips T}$$

Example 10-2. With reference to the frame in Fig. 10-6a, find the reactions at the supports and the forces in members BF and BC.

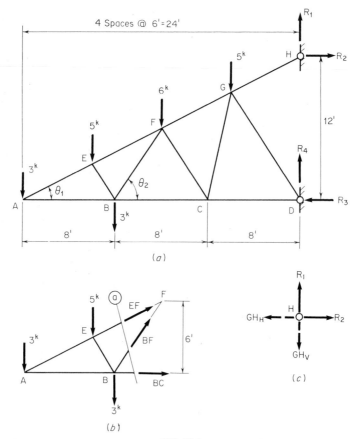

FIG. 10-6

Solution. We shall first find the forces in the specified members. Cut the frame along the plane *a* shown in Fig. 10-6*b*. Resolve *BF* into its horizontal and vertical components, performing the resolution at *B*.

$$\Sigma M_A = 5 \times 6 + 3 \times 8 - 8BF_V = 0$$
$$\therefore BF_V = 6.75 \text{ kips}$$

In Fig. 10-6*a*, $\tan \theta_2 = 6/4 = 1.500$ $\therefore \sin \theta_2 = 0.832$

$$BF = \frac{6.75}{0.832} = \textbf{8.11 kips T}$$

$$\Sigma M_F = -3 \times 12 - 5 \times 6 - 3 \times 4 - 6BC = 0$$
$$\therefore BC = -13 \text{ kips} \qquad \text{or} \qquad BC = \textbf{13 kips C}$$

We shall now calculate the reactions. In Fig. 10-6*a*,

$$\tan \theta_1 = 12/24 = 0.500$$
$$\Sigma M_D = -3 \times 24 - 5 \times 18 - 6 \times 12 - 5 \times 6 - 3 \times 16 + 12R_2 = 0$$
$$R_2 = 26 \text{ kips to right} \qquad \text{and} \qquad R_3 = 26 \text{ kips to left}$$

Resolve *GH* into its horizontal and vertical components at *H*, as shown in Fig. 10-6*c*. At this joint,

$$\Sigma F_H = R_2 - GH_H = 26 - GH_H = 0 \qquad GH_H = 26 \text{ kips}$$
$$GH_V = GH_H \tan \theta_1 = 13 \text{ kips} \qquad \therefore R_1 = 13 \text{ kips}$$

For the frame,

$$\Sigma F_V = R_1 + R_4 - (3 + 5 + 6 + 5 + 3) = 0$$
$$\therefore R_4 = 9 \text{ kips}$$

10-4   Graphical Method.

In design practice, it is necessary to determine the forces in all truss members, and the graphical method of investigation commends itself for this purpose. It is generally simple, rapid, and reasonably accurate. The method consists of cutting the truss at every joint and then constructing the force polygon corresponding to the system of forces acting at each joint. The polygons are juxtaposed so that every vector representing an internal force forms the side of two adjacent polygons. In this manner, the duplication of lines is avoided, and there evolves a composite force polygon corresponding to the truss itself.

Example 10-3. Determine graphically the forces in the truss members in Fig. 10-7*a*.

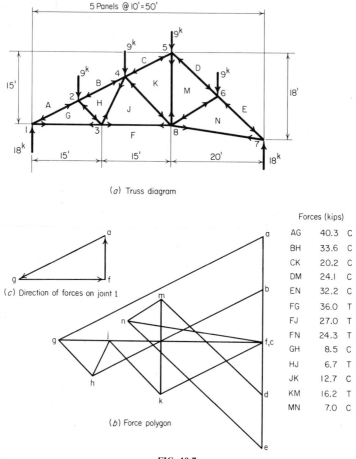

(a) Truss diagram

Forces (kips)

AG	40.3	C
BH	33.6	C
CK	20.2	C
DM	24.1	C
EN	32.2	C
FG	36.0	T
FJ	27.0	T
FN	24.3	T
GH	8.5	C
HJ	6.7	T
JK	12.7	C
KM	16.2	T
MN	7.0	C

(c) Direction of forces on joint 1

(b) Force polygon

FIG. 10-7

Solution. This drawing applies Bow's notation, in which every region of space bounded by the action lines of either external or internal forces is assigned an upper-case letter and the forces and their corresponding members are designated by the regions they delimit. For example, the upper-chord member at the extreme left is *AG*, the load at the apex is *CD*, and the vertical member is *KM*. In the force polygon, the forces are represented by the corresponding lower-case letters. The joints have been numbered in the manner shown.

Using a suitable scale, draw the load line *abcde* in Fig. 10-7*b* to represent the loads *AB*, *BC*, *CD*, and *DE*. Now locate point *f* so that *ef* and *fa* represent the reactions *EF* and *FA*, respectively. Since both reactions are 18 kips, *f* is the midpoint of *ae*.

Consider the truss to be cut at joint 1. Three forces act at this joint: *AG*, *GF*, and *FA*. In Fig. 10-7*b*, draw lines through *a* and *f* parallel, respectively, to the upper and lower chords, these lines intersecting at *g*. Vectors *ag* and *gf* represent, respectively, the forces *AG* and *GF* at joint 1. Refer to Fig. 10-7*c*, which is a reproduction of the force polygon at joint 1, but to a reduced scale. Since the reaction *FA* is directed upward, it follows that *AG* is directed toward the joint and *GF* away from the joint. The sense of these forces is indicated by the arrowheads in Fig. 10-7*a*.

With *AG* determined, there exist only two unknown forces at joint 2. Therefore, proceed to this joint, construct its force polygon *abhg*, determine the sense of each force, and record the sense in the truss diagram. After joint 2 is analyzed, proceed to joint 3, where there now exist only two unknown forces. Then proceed to joints 4, 5, and 6, in that order.

The construction of the force polygon at joint 6 serves to complete the composite force polygon. To verify the accuracy of the construction, ascertain that line *fn* in the force polygon is parallel to member *FN* in the truss diagram. Now scale the vectors in the force polygon to secure the magnitudes of the internal forces; these are tabulated in the drawing. Following the procedure illustrated in Fig. 10-7*c*, establish the direction of each force acting on a joint to determine the character of the internal forces. A truss member is in tension or compression according to whether its force is directed away from or toward the joint, respectively.

10-5 Considerations in Design of Roof Truss. The design of a roof truss has certain special features. The purlins framing to the upper chord of the truss transmit to the structure the snow load, weight of roof, and wind load. In many instances, the truss also supports ceiling, machinery, or other loads transmitted to the structure by beams framing to the lower chord.

When a structure is designed for a combination of snow and wind loads, the AISC Specification stipulates that the allowable stresses are to be increased by one-third. This provision is based on the fact that the coexistence of maximum snow and maximum wind load is improbable. Therefore, since the allowable stresses depend on the nature of the load system, a roof truss must be analyzed for two distinct sets of conditions: gravity loads acting alone and gravity loads acting in conjunction with wind loads. Moreover, if the truss or the gravity-load system has dissymmetry, three possible load systems must be taken into account, for there are two possible directions of the wind.

10-6   Loading of Bridge Truss.   The floor of a bridge is carried by transverse beams that frame to the two supporting trusses at their joints, as shown in Fig. 10-8. In a *through* bridge, the floor beams are connected to the lower chords; in a *deck* bridge, to the upper chords. The points at which the floor beams frame to a bridge truss are termed its *panel points.* Thus, every load on the bridge is transmitted to the supporting trusses in the form of concentrated loads acting at their panel points. In the subsequent material, it is to be understood that the specified magnitude of a load is that part of the load that is transmitted to the truss under consideration.

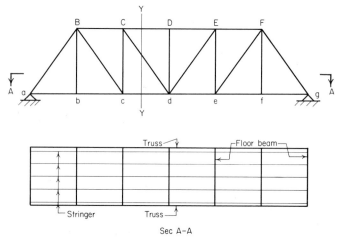

FIG. 10-8. Pratt truss and bridge floor.

A bridge truss carries stationary dead loads and moving live loads. Therefore, to calculate the maximum internal force that will be induced in a given truss member during transit of the live-load system, it is first necessary to identify the disposition of the system that will induce this maximum internal force. It follows that the cardinal problem in the design of a bridge truss is to identify the critical live-load dispositions, and we shall develop an analytical tool that aids in solving this problem.

10-7 Vertical Shear and Bending Moment in Bridge Truss. With reference to the through-bridge truss in Fig. 10-8, assume that we wish to find the force in member Cd caused by a given load system. To do this, we cut the truss along plane Y and consider the equilibrium of that part of the truss lying to the left of this plane. The force in Cd is found by applying the equation $\Sigma F_V = 0$. Therefore, the force in Cd is a function of the algebraic sum of the forces to the left of plane Y. By analogy with the nomenclature pertaining to a beam, this quantity is called the *vertical shear* in panel cd. We therefore have the following definition:

The vertical shear in a given panel is the algebraic sum of the vertical forces to the left of a plane through that panel. It is denoted by V.

Now assume that we wish to find the force in member CD in Fig. 10-8. We again cut the truss along plane Y and consider the equilibrium of the left part of the truss. The force in CD is found by taking moments with respect to d, thereby securing an equation containing force CD as the sole unknown. Therefore, this force is a function of the algebraic sum of the moments of all forces to the left of d with respect to d. This quantity is called the *bending moment* at d, and we now have the following definition:

The bending moment at a given panel point is the algebraic sum of the moments of all forces to the left of that point with respect to that point. It is denoted by M.

The sign convention for vertical shear and bending moment in a truss is identical with that given in Art. 1-3. In calculating the bending moment at d, it is permissible to include the load at d if convenient, since the moment of that load with respect to d is zero.

In general, the total load to be transmitted from the bridge floor to the given truss is resolved by the stringers and floor beams into a group of loads acting at the panel points. These panel-point loads constitute a force system that is equivalent to the total load. Therefore, if all panel-point loads would be included in a given calculation, we may apply the total load itself for simplicity. As an illustration, assume that a load of 8 kips is transmitted to the truss in Fig. 10-8. Assume also that this load lies in panel bc, inducing a load of 5 kips at b and 3 kips at c. Since both panel-point loads enter into the calculation of the reaction at a and the vertical shear in panel cd, we may apply the 8-kip load directly in these calculations. On the other hand, we must consider the panel-point loads rather than the total load in calculating the vertical shear in panel *bc*.

10-8   Influence Lines.   Figure 10-9*a* shows the lower chord of a through-bridge truss of eight panels and the floor beams that it supports. Consider that a unit concentrated load traverses the bridge. Locomotion is assumed to proceed from right to left, and *x* denotes the distance from the right end of the truss to the load. Every variable quantity associated with this moving load, such as the load at a panel point, the reaction at a support, the vertical shear in a panel, and the bending moment at a panel point, is a function

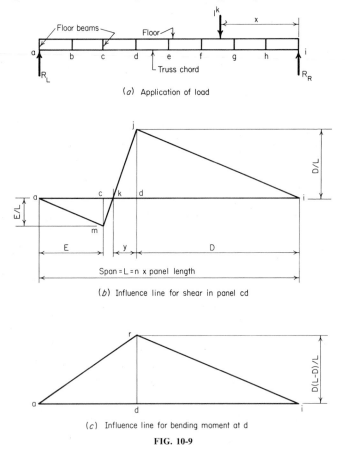

(*a*) Application of load

(*b*) Influence line for shear in panel cd

(*c*) Influence line for bending moment at d

FIG. 10-9

of x; i.e., its value depends on the instantaneous position of the unit load.

An *influence line* is a diagram that depicts the variation of a given quantity with x. Its base line corresponds to the loaded chord of the truss, and the ordinate at a given point represents the value of the variable quantity when the unit load is at the given point. In general, therefore, if the bridge carries a concentrated moving load of magnitude P, the given quantity is evaluated by multiplying the ordinate of the influence line by P. If the bridge carries a uniformly distributed load of intensity w across a given interval, the given quantity is evaluated by calculating the area under the influence line across that interval and multiplying this area by w.

Figure 10-9*b* is the influence line for vertical shear in panel *cd*. Thus, when the unit load is at d, $V = dj$; when the unit load is at c, $V = -cm$. In addition to the notation shown in the drawing, let

r_c = panel-point load at c
n = number of panels
p = panel length

The equations of the influence line are these: With load to right of *cd*:

$$V = R_L = \frac{x}{L} \tag{a}$$

With load to left of *cd*:

$$V = R_L - 1 = \frac{x}{L} - 1 \tag{b}$$

With load within *cd*:

$$V = R_L - r_c = \frac{x}{L} - \frac{x - D}{p} = \frac{x}{np} - \frac{x - D}{p}$$

or

$$V = \frac{D}{p} - \frac{(n - 1)x}{np} \tag{c}$$

The point k at which the influence line intersects the base is called the *neutral point*. The location of this point is found by setting $V = 0$ in Eq. (*c*), giving

$$x = \frac{n}{n - 1} D \quad \text{and} \quad y = \frac{D}{n - 1} \tag{10-1}$$

Consider that the bridge sustains a moving concentrated load P. The influence line reveals that the extreme values of vertical shear in cd are

$$V_{max} = \frac{PD}{L} \quad \text{and} \quad V_{min} = -\frac{PE}{L} \quad (10\text{-}2)$$

Now consider that the bridge sustains a uniformly distributed live load of intensity w. The vertical shear in cd is maximum when the load extends continuously from i to k, and minimum when it extends continuously from k to a. Multiplying the area of each triangle in Fig. 10-9b by w yields

$$V_{max} = \frac{wD^2}{2(D + E)} \quad \text{and} \quad V_{min} = -\frac{wE^2}{2(D + E)} \quad (10\text{-}3)$$

Since the vertical shear assumes both positive and negative values, member Cd undergoes a reversal of stress during transit of the load. Both the maximum tensile force and the maximum compressive force in the member must be evaluated.

Figure 10-9c is the influence line for bending moment at d. A concentrated load P induces the maximum bending moment when it lies directly above d, and

$$M_{max} = \frac{PD(L - D)}{L} \quad (10\text{-}4)$$

A uniform live load of intensity w induces maximum bending moment when it extends across the entire span, and

$$M_{max} = \frac{wD(L - D)}{2} \quad (10\text{-}5)$$

It is apparent that the use of influence lines constitutes a highly effective semigraphical method of analysis and that it readily reveals the critical dispositions of live load.

Example 10-4.   With reference to the through Pratt truss in Fig. 10-8, the panel length is 20 ft and the height of truss is 25 ft. The truss sustains an equivalent uniform live load of 2 klf and a concentrated live load of 10 kips. Compute the maximum force due to live load in Bb, BC, and Cd.

Solution.   Force Bb equals the load transmitted by the floor beam at b. To maximize this load, place the 10-kip load at b and apply uniform load from c to a.

$$Bb = 10 + \tfrac{1}{2} \times 2 \times 40 = \textbf{50 kips T}$$

To determine BC, cut the truss in the second panel. Force BC is a function of the bending moment at c. We shall first calculate this bending moment without recourse to Eqs. (10-4) and (10-5). Place the 10-kip load at c, and apply uniform load to the entire span.

$$R_L = 10 \times \tfrac{4}{6} + \tfrac{1}{2} \times 2 \times 120 = 126.7 \text{ kips}$$

The total uniform load from c to a is 2×40 or 80 kips. For the force system to the left of the cutting plane,

$$\Sigma M_c = 126.7 \times 40 - 80 \times 20 + 25BC = 0$$
$$\therefore BC = \textbf{138.7 kips C}$$

To obtain this result by formula, combine Eqs. (10-4) and (10-5).

$$M_c = \frac{10 \times 80 \times 40}{120} + \frac{2 \times 80 \times 40}{2} = 3467 \text{ ft-kips}$$

Then
$$BC = \tfrac{3467}{25} = 138.7 \text{ kips C}$$

To determine Cd, cut the truss in the third panel and compute the vertical shear in this panel. We shall first do this without recourse to Eqs. (10-2) and (10-3). For maximum vertical shear, place the 10-kip load at d, and apply uniform load from g to the neutral point for that panel.

$$y = \frac{D}{n-1} = \frac{60}{5} = 12 \text{ ft}$$
$$R_L = \tfrac{1}{2} \times 10 + 2 \times 72 \times \tfrac{36}{120} = 48.2 \text{ kips}$$

The panel-point load at c is

$$r_c = 2 \times 12 \times \tfrac{6}{20} = 7.2 \text{ kips}$$
$$V_{\max} = 48.2 - 7.2 = 41.0 \text{ kips}$$
Then
$$Cd_V = 41.0 \text{ kips} \qquad \therefore Cd = \textbf{52.5 kips T}$$

For minimum vertical shear, place the 10-kip load at c and apply uniform load from the neutral point to a.

$$R_L = 10 \times \tfrac{4}{6} + 2 \times 48 \times \tfrac{96}{120} = 83.5 \text{ kips}$$

The panel-point load at c due to the uniform load in the third panel is

$$r_c' = 2 \times 8 \times \tfrac{16}{20} = 12.8 \text{ kips}$$
$$V_{\min} = 83.5 - 2 \times 40 - 12.8 - 10 = -19.3 \text{ kips}$$
Then
$$Cd_V = -19.3 \text{ kips} \qquad \therefore Cd = \textbf{24.7 kips C}$$

The vertical-shear values may also be obtained by combining Eqs. (10-2) and (10-3), giving

$$V_{max} = \frac{10 \times 60}{120} + \frac{2 \times 60^2}{2(60 + 40)} = 41.0 \text{ kips}$$

$$V_{min} = -\frac{10 \times 40}{120} - \frac{2 \times 40^2}{2(60 + 40)} = -19.3 \text{ kips}$$

The calculation of vertical shear by formula is clearly more expedient.

With reference to the deck-bridge truss in Fig. 10-10a, assume that we wish to determine the live-load forces in members Bb and

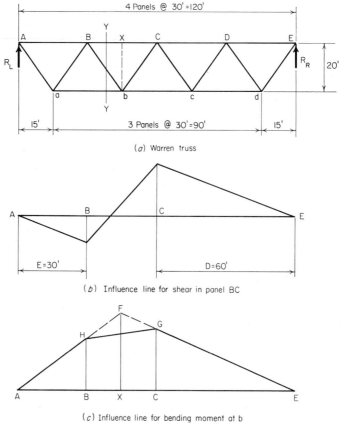

(*a*) Warren truss

(*b*) Influence line for shear in panel BC

(*c*) Influence line for bending moment at b

FIG. 10-10

BC. Cut the truss along plane *Y*. Force *Bb* is a function of the vertical shear in panel *BC*, and the influence line for this quantity appears in Fig. 10-10*b*. Force *BC* is a function of the bending moment at *b* (or at *X*), and the influence line for this quantity consists of the solid lines in Fig. 10-10*c*. Since this diagram is a quadrilateral, we shall not formulate an equation for the maximum bending moment, as it would be too cumbersome to be useful.

Example 10-5. The truss in Fig. 10-10*a* carries an equivalent uniform live load of 2 klf and a concentrated live load of 15 kips. Compute the maximum forces in *Bb* and *BC*.

Solution. In calculating the shear in panel *BC*, Eqs. (10-2) and (10-3) are applicable, and they yield the following:

$$V_{max} = \frac{15 \times 60}{120} + \frac{2 \times 60^2}{2(60 + 30)} = 47.5 \text{ kips}$$

$$Bb = 1.25 \times 47.5 = \textbf{59.4 kips T}$$

$$V_{min} = -\frac{15 \times 30}{120} - \frac{2 \times 30^2}{2(60 + 30)} = -13.8 \text{ kips}$$

$$Bb = 1.25 \times 13.8 = \textbf{17.3 kips C}$$

For the maximum force in *BC*, place the 15-kip load at *C* and apply uniform load to the entire span.

$$R_L = \tfrac{1}{2} \times 15 + \tfrac{1}{2} \times 2 \times 120 = 127.5 \text{ kips}$$

The panel-point loads at *A* and *B* are

$$r_A = 30 \text{ kips} \qquad \text{and} \qquad r_B = 60 \text{ kips}$$

For the force system to the left of the cutting plane,

$$\Sigma M_b = 127.5 \times 45 - 30 \times 45 - 60 \times 15 + 20BC = 0$$
$$\therefore BC = \textbf{174.4 kips C}$$

10-9 Forces in Bridge Truss under Multiple-Load Systems. We shall now analyze the internal forces of a bridge truss that result from transit of a group of concentrated live loads. For this purpose, we shall refer to Fig. 10-9 and investigate the vertical shear in panel *cd* and the bending moment at *d*. Let

W = sum of loads on span
Q = sum of loads in panel *cd*
T = sum of loads to right of *d*

Figure 10-9*b* discloses that as a single concentrated unit load advances from right to left *V* varies in this manner: It increases at the rate of 1/*L* when the load lies outside the given panel and it decreases at the rate of $(n-1)/L$ when the load lies within this panel. Therefore, with respect to the multiple-load system,

$$\frac{dV}{dx} = \frac{W - Q}{L} - \frac{Q(n-1)}{L} = \frac{W - Qn}{L}$$

It follows that *V* is increasing when *Q* is less than *W*/*n* and decreasing when *Q* exceeds *W*/*n*. Therefore, *V* is maximum when a transition from the first state to the second state impends, and minimum when a transition in the reverse direction impends.

Similarly, Fig. 10-9*c* yields

$$\frac{dM}{dx} = \frac{T(L - D)}{L} - \frac{(W - T)D}{L} = \frac{TL - WD}{L}$$

It follows that *M* is increasing when *T* exceeds *WD*/*L* and decreasing when *T* is less than *WD*/*L*. Therefore, *M* is maximum when a transition from the first state to the second state impends. In summary,

$$\begin{array}{ll}
\text{Before } V \text{ is maximum} \dots & Q < W/n \\
\text{After } V \text{ is maximum} \dots & Q > W/n \\
\text{Before } M \text{ is maximum} \dots & T > WD/L \\
\text{After } M \text{ is maximum} \dots & T < WD/L
\end{array}$$

The foregoing relationships afford criteria for identifying the critical disposition of the multiple-load system.

Example 10-6.   The Pratt truss in Fig. 10-11*a* carries the indicated group of live loads at its bottom chord. Compute the maximum forces in *Bb* and *CD*, and the maximum tensile force in *Bc*.

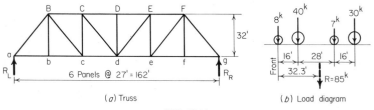

(*a*) Truss   (*b*) Load diagram

FIG. 10-11

Solution.   The magnitude and location of the resultant of the group are recorded in the load diagram. Force Bb equals the panel-point load r_b at b. To maximize this load, place the 40-kip load directly above b. The 8-kip load then falls in panel ab, and the two remaining loads fall to the right of d. Then

$$Bb = r_b = 40 + 8 \times \tfrac{11}{27} = \textbf{43.3 kips T}$$

Cut the truss in the third panel. Force CD is a function of the bending moment at d. With respect to d,

$$\frac{WD}{L} = \frac{85 \times 81}{162} = 42.5 \text{ kips}$$

Now, $40 + 7 + 30 > 42.5$ but $7 + 30 < 42.5$

Therefore, the bending moment at d attains its maximum value when the 40-kip load lies above d. Simultaneously, the resultant lies 64.7 ft from g. Then

$$R_L = 85 \times 64.7/162 = 33.9 \text{ kips}$$

For the load system to the left of the cutting plane,

$$\Sigma M_d = 33.9 \times 81 - 8 \times 16 + 32CD = 0$$
$$\therefore CD = \textbf{81.8 kips C}$$

Alternatively, the force CD can be found by this procedure: Construct the influence line for bending moment at d, position the load group in the prescribed manner, multiply each load in the group by the ordinate to the influence line at that location, sum the results to obtain the bending moment at d, and divide this moment by 32.

Cut the truss in the second panel. Force Bc is a function of the vertical shear in this panel. By convention, it is assumed that bridge traffic flows solely from right to left, but in reality it flows in both directions. In the present instance, it is not evident which direction of motion induces the maximum vertical shear in panel bc.

This problem can be resolved by considering the vertical shear in panel ef, the counterpart of bc on the opposite side of the truss. Figure 10-12 presents the influence lines for these two panels. If either diagram is revolved about the horizontal base line, the two diagrams become opposite hand to one another. Consider first that a dual-load system is traversing the bridge from left to right and has the instantaneous location shown in Fig. 10-12a. Now consider that the system is traversing the bridge from right to left and has the instantaneous location shown in Fig. 10-12b. It follows from the symmetry of the drawings that the vertical shear in bc in the first case is numerically equal to the vertical shear in ef in the second case. Therefore, the maximum

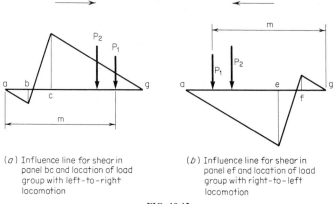

(*a*) Influence line for shear in panel bc and location of load group with left-to-right locomotion

(*b*) Influence line for shear in panel ef and location of load group with right-to-left locomotion

FIG. 10-12

positive shear in panel *bc* with left-to-right locomotion equals the maximum negative (i.e., minimum) shear in panel *ef* with right-to-left locomotion. This principle enables us to assume a single-track movement and calculate all internal forces accordingly. Since in design practice it is necessary to determine all maximum and minimum panel shears, the ability to assume single-track movement affords enormous economy of effort.

For this truss,

$$\frac{W}{n} = \frac{85}{6} = 14.2 \text{ kips}$$

With respect to panel *bc*, a transition from $Q < W/n$ to $Q > W/n$ occurs twice: when the 40-kip load is on the verge of entering the panel, and later when the 30-kip load is on the verge of entering. The influence line clearly indicates that the vertical shear in panel *bc* attains its maximum value when the 40-kip load lies at *c*. Simultaneously, the resultant lies 91.7 ft from *g*. Then

$$R_L = 85 \times 91.7/162 = 48.1 \text{ kips}$$

The panel-point load at *b* is

$$r_b = 8 \times \tfrac{16}{27} = 4.7 \text{ kips}$$

Then

$$V_{bc,\text{max}} = 48.1 - 4.7 = 43.4 \text{ kips}$$

Similarly, with respect to panel *ef*, a transition from $Q > W/n$ to $Q < W/n$ occurs twice: when the 40-kip load is on the verge of leaving the panel, and later when the 30-kip load is on the verge of leaving. The influence line clearly indicates that the vertical shear in *ef* attains its minimum value when the

30-kip load lies at *e*. Then

$$R_L = 85 \times 81.7/162 = 42.9 \text{ kips}$$

and $$V_{ef,\min} = 42.9 - 85 = -42.1 \text{ kips}$$

Comparing the two results, we find that the maximum positive shear in panel *bc* with both directions of motion taken into account is 43.4 kips. Then

$$Bc_V = 43.4 \text{ kips} \qquad \text{and} \qquad Bc = 1.3 \times 43.4 = \textbf{56.4 kips T}$$

PROBLEMS

10-1. With reference to the frame in Fig. 10-6*a*, find the force in *BE*.

ANS. 4.51 kips C

10-2. With reference to the frame in Fig. 10-6*a*, find the forces in *AE*, *CD*, *FG*, and *CG*.

ANS. $AE = 6.71$ kips T, $CD = 20$ kips C, $FG = 20.40$ kips T, $CG = 8.07$ kips T

10-3. With reference to the truss in Fig. 10-13, calculate the horizontal and vertical components of the forces in *AB* and *EF*, and the force in *CF*. Indicate the character of each force. Find the force in *CF* by each of these methods:

a. Cut the truss through *EF*, *CF*, and *CD*, and consider the equilibrium of the left part of the truss. Set $\Sigma F_V = 0$, and apply the value of EF_V previously calculated.

b. Cut the truss in the same manner, and take moments with respect to the point of intersection of *EF* and *CD* prolonged.

ANS. $AB_H = 1939$ lb, $AB_V = 776$ lb, tension; $EF_H = 916$ lb, $EF_V = 305$ lb, compression; $CF = 535$ lb T

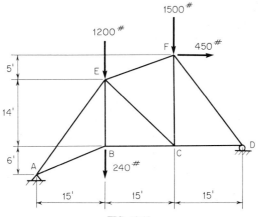

FIG. 10-13

10-4. The deck-bridge truss in Fig. 10-10*a* carries an equivalent uniform live load of 600 plf and the live-load group shown in Fig. 10-14. Determine the following:

    a. The maximum shear in panel *AB* with right-to-left locomotion.

                                                             ANS. 61.2 kips

    b. The maximum shear in panel *AB* with left-to-right locomotion.

                                                             ANS. 59.6 kips

    c. The maximum force in *aB*.   ANS. 76.5 kips C

    d. The maximum bending moment at *C*.   ANS. 2420 ft-kips

    e. The maximum force in *bc*.   ANS. 121 kips T

    f. The maximum bending moment at *b* caused solely by the concentrated-load group with right-to-left locomotion, and the corresponding disposition of the group.

                                    ANS. 1097 ft-kips; 15-kip load at C

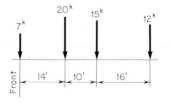

FIG. 10-14

10-5. The deck-bridge truss in Fig. 10-10*a* carries the group of moving concentrated loads shown in Fig. 10-15. Prove that when the group is moving from left to right the bending moment at *b* remains constant from the instant the 12-kip load arrives at *C* to the instant the 6-kip load arrives at *B*.

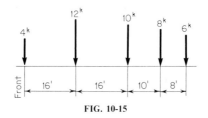

FIG. 10-15

11

TIMBER DESIGN

11-1   Characteristics of Wood.   By virtue of its fibrous composition, wood is an *aeolotropic* rather than an isotropic material; that is, the mechanical properties of a wood body are dependent upon the direction of the applied forces. A timber member subjected to direct forces exhibits its maximum strength when the direction of the forces is parallel to the grain, and its minimum strength when the direction is normal to the grain. Consequently, the optimum exploitation of wood as a structural material hinges on its proper orientation relative to the external forces.

In expressing the size of a timber member, it is necessary to distinguish between the *nominal* and *true* dimensions. The nominal dimensions, which are used for identification purposes, pertain to the rough, undressed member; the true dimensions pertain to the finished product. To convert a nominal dimension to a true dimension, deduct $\frac{3}{8}$ in. if the former is less than 6 in., and deduct $\frac{1}{2}$ in. if it is 6 in. or more. Thus, a "4 by 8" section is actually $3\frac{5}{8}$ by $7\frac{1}{2}$ in.

The weight of timber may be taken as 40 pcf in the absence of any statement to the contrary.

11-2   Flexural Members.   The capacity of a wood beam is often controlled by shear or deflection rather than bending. In computing the maximum vertical shear in the member, it is permissible to disregard the load that lies within a distance equal to the beam depth from the ends, but we shall include the entire load for simplicity. The maximum shearing stress on a rectangular section is given by Eq. (5-17).

Example 11-1.   A floor is supported by 3- by 8-in. wood joists spaced 16 in. on centers and having an effective span of 11 ft. What load can the floor carry if the allowable stresses in bending and horizontal shear, respectively, are 1200 and 100 psi, the deflection is limited to 0.4 in., and E is 1,000,000 psi?

Solution

$$A = 19.7 \text{ sq in.} \qquad\qquad I = 92.3 \text{ in.}^4$$
$$S = 24.6 \text{ in.}^3 \qquad \text{Beam weight} = 5 \text{ plf}$$

Let w denote the load per linear foot on a joist.

$$M = \tfrac{1}{8}w \times 11^2 \times 12 = fS = 1200 \times 24.6$$

$\therefore w = 163$ plf if flexural stress controls.

$$V = \tfrac{1}{2} \times 11w = \tfrac{2}{3}Av = \tfrac{2}{3} \times 19.7 \times 100$$

$\therefore w = 239$ plf if shearing stress controls.

$$y = \frac{5}{384}\frac{wL^4}{EI} = \frac{5}{384}\frac{w \times 11^4 \times 1728}{1,000,000 \times 92.3} = 0.4$$

$\therefore w = 112$ plf if deflection controls.

$$\text{Net allowable load} = 112 - 5 = 107 \text{ plf}$$
$$\text{Allowable floor load} = 107/1.33 = \textbf{80 psf}$$

This load includes the weight of the wood planks.

Example 11-2.   A concrete platform weighing 100 psf is to be supported during pouring by a timber deck consisting of 1-in. planks (exact size) resting on wood joists. Calculate the allowable spacing of joists, considering the plank to be simply supported. The allowable stresses in flexure and shear, respectively, are 1500 and 100 psi, E is 1,500,000 psi, and the allowable deflection of the plank is $\tfrac{1}{300}$ of the span. This timber weighs 48 pcf.

Solution. We shall consider a strip of plank 1 ft wide. The plank weighs 4 psf, and the total load on the strip is therefore 104 plf.

$$A = 1 \times 12 = 12 \text{ sq in.}$$

$$I = \tfrac{1}{12} \times 12 \times 1^3 = 1 \text{ in.}^4 \qquad S = \frac{1}{\frac{1}{2}} = 2 \text{ in.}^3$$

In Fig. 11-1, let L denote the center-to-center spacing of joists in feet, and consider this as the effective span.

$$M = \tfrac{1}{8} \times 104 \times L^2 \times 12 = fS = 1500 \times 2$$

$\therefore L = 4.4$ ft if flexural stress controls .

$$V = \tfrac{1}{2} \times 104 \times L = \tfrac{2}{3}Av = \tfrac{2}{3} \times 12 \times 100$$

$\therefore L = 15.4$ ft if shearing stress controls.

$$y = \frac{5}{384} \frac{104 \times L^4 \times 1728}{1,500,000 \times 1} = \frac{L \times 12}{300}$$

$\therefore L = 2.9$ ft if deflection controls. Use **3-ft** spacing of joists.

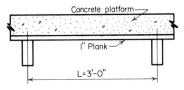

Concrete platform

1" Plank

L = 3'-0"

FIG. 11-1

11-3 Compression on Oblique Plane.

Consider that a timber member sustains a compressive force whose action line makes an oblique angle with the grain. Let

   P = allowable compressive stress parallel to grain
   Q = allowable compressive stress normal to grain
   N = allowable compressive stress inclined to grain
   θ = angle between action line of N and direction of grain

Then
$$\frac{N \cos^2 \theta}{P} + \frac{N \sin^2 \theta}{Q} = 1$$

or
$$N = \frac{PQ}{P \sin^2 \theta + Q \cos^2 \theta} \tag{11-1}$$

The latter relationship is known as Hankinson's formula.

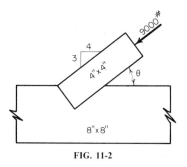

FIG. 11-2

Example 11-3. If the allowable compressive stresses are 1400 and 400 psi parallel and normal to the grain, respectively, is the joint in Fig. 11-2 satisfactory with respect to bearing?

Solution

$$P = 1400 \text{ psi} \qquad Q = 400 \text{ psi}$$

$$f = \frac{9000}{3.625^2} = 685 \text{ psi}$$

$$\sin^2 \theta = 3^2/5^2 = 0.36 \qquad \cos^2 \theta = 4^2/5^2 = 0.64$$

$$N = \frac{1400 \times 400}{1400 \times 0.36 + 400 \times 0.64} = 737 \text{ psi} > f$$

∴ Joint is satisfactory.

In Fig. 11-3, member $M1$ must be notched at the joint to avoid removing an excessive area from member $M2$. If the member is cut in such a manner that AC and BC make an angle of $\phi/2$ with vertical and horizontal planes, respectively, the allowable bearing pressures at these faces are identical for the two members. If A denotes the sectional area of member $M1$, f_1 the pressure at AC,

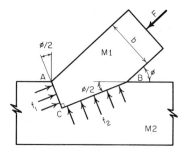

FIG. 11-3. Half-angle notch.

and f_2 the pressure at BC, it can be readily shown that

$$AC = b\,\frac{\sin\,(\phi/2)}{\sin\,\phi} \qquad BC = b\,\frac{\cos\,(\phi/2)}{\sin\,\phi} \qquad (11\text{-}2)$$

$$f_1 = \frac{F\,\sin\,\phi}{A\,\tan\,(\phi/2)} \qquad f_2 = \frac{F\,\sin\,\phi\,\tan\,(\phi/2)}{A} \qquad (11\text{-}3)$$

This type of joint is often used in wood trusses.

Example 11-4. With reference to Fig. 11-3, design a suitable notch using the following data: $M1$ is a 4 by 4, F is 5500 lb, ϕ is 30°, P is 1200 psi, and Q is 390 psi. The cut in $M2$ is restricted to $2\frac{1}{2}$ in. vertically.

Solution. We shall try a half-angle notch.

$$\sin 30° = 0.500 \qquad \cos 15° = 0.966$$
$$\sin 15° = 0.259 \qquad \tan 15° = 0.268$$

The area of a 4 by 4 is 13.1 sq in.

$$f_1 = \frac{5500 \times 0.500}{13.1 \times 0.268} = 783 \text{ psi}$$

$$f_2 = \frac{5500 \times 0.500 \times 0.268}{13.1} = 56 \text{ psi}$$

$$N_1 = \frac{1200 \times 390}{1200 \times 0.259^2 + 390 \times 0.966^2} = 1053 \text{ psi} \qquad \text{OK}$$

$$N_2 = \frac{1200 \times 390}{1200 \times 0.966^2 + 390 \times 0.259^2} = 408 \text{ psi} \qquad \text{OK}$$

$$AC = 3.625 \times 0.259/0.500 = 1.9 \text{ in.}$$
$$BC = 3.625 \times 0.966/0.500 = 7.0 \text{ in.}$$

Make AC 2 in. and BC 7 in.

11-4 Timber Columns. For the design of timber columns, we shall apply the recommendations of the Forest Products Laboratory. Let

P = allowable load on solid column

A = cross-sectional area

L = unbraced length

d = smaller side of rectangular section

E = modulus of elasticity

f_c = allowable compressive stress parallel to grain when buckling tendency is absent

Timber columns are divided into two categories: long columns, which tend to buckle at the stress given by Euler's column formula, and intermediate columns, which tend to fail at a higher stress. For a rectangular section, the minimum radius of gyration is $d/2 \sqrt{3}$, and it is convenient to deal with the L/d ratio directly.

By setting the allowable stress equal to one-third the compressive stress at which buckling occurs in accordance with Euler's formula, we obtain the following equation for a long column:

$$\frac{P}{A} = \frac{0.274E}{(L/d)^2} \tag{11-4}$$

Let K denote the value of L/d at which the column will buckle when the compressive stress reaches $\frac{2}{3}f_c$. Then

$$K = 0.64 \left(\frac{E}{f_c}\right)^{1/2} \tag{11-5}$$

An intermediate column is one for which $L/d \leqslant K$, and its allowable stress is

$$\frac{P}{A} = f_c \left[1 - \frac{1}{3}\left(\frac{L/d}{K}\right)^4 \right] \tag{11-6}$$

The allowables stress in an intermediate column therefore lies between f_c and $\frac{2}{3}f_c$.

Example 11-5.   An 8- by 10-in. timber column has an unbraced length of 10 ft 6 in. What is the capacity of this member if f_c is 1500 psi and E is 1,760,000 psi?

Solution

$$A = 71.3 \text{ sq in.} \qquad \frac{L}{d} = \frac{126}{7.5} = 16.8$$

$$K = 0.64 \left(\frac{1,760,000}{1500}\right)^{1/2} = 21.9$$

This is an intermediate column.

$$P = 71.3 \times 1500 \left[1 - \frac{1}{3}\left(\frac{16.8}{21.9}\right)^4 \right] = \textbf{94, 600 lb}$$

Example 11-6. A timber column with an unbraced length of 12 ft sustains a load of 98 kips. Design a solid section, using $f_c = 1400$ psi and $E = 1,760,000$ psi.

Solution. Assume that the smaller side of the section is $7\frac{1}{2}$ in.

$$K = 0.64 \left(\frac{1,760,000}{1400}\right)^{1/2} = 22.7$$

$$\frac{L}{d} = \frac{144}{7.5} = 19.2$$

This is an intermediate column if our assumption is correct.

$$\text{Allowable stress} = 1400 \left[1 - \frac{1}{3}\left(\frac{19.2}{22.7}\right)^4\right] = 1161 \text{ psi}$$

$$\text{Area required} = 98,000/1161 = 84.4 \text{ sq in.}$$

∴ Use an **8- by 12-in.** section, which has an area of 86.3 sq in.

12

LOADS ON FOUNDATIONS—
STABILITY OF STRUCTURES

12-1 Determination of Soil Pressure. Assume that a structure is supported by a homogeneous stratum of compressible soil having horizontal surfaces at top and bottom and that this soil in turn is supported by incompressible bedrock. Assume also that the structure has a rigid base and that it sustains solely a vertical load, including its own weight, that is concentric with respect to the base. This load is transmitted to the underlying soil, but the manner in which it is distributed is rather complex. For simplicity, it is often assumed that the soil pressure (load on a unit area) is uniform directly below the base.

Now consider that a moment is applied to the structure. This moment is transmitted to the base, causing it to rotate slightly about its centroidal axis normal to the plane of the moment. Since the surface of contact between the base and the underlying soil remains a plane, it may be assumed that the soil pressure varies uniformly from toe to heel. By analogy with solid structural members, the soil pressure directly below the base can be calculated by assuming that the axial load and moment on the structure are resisted solely by the vertical prism of soil that underlies the base. This prism of soil thus functions as a member that is subjected simultaneously to an axial compressive load and flexure.

Example 12-1. A pedestal-type footing weighing 50 kips supports a column with a load of 200 kips. At the base of the column, which is 5 ft above the base of the footing, there is a longitudinal horizontal force of 15 kips and a transverse horizontal force of 10 kips. The footing base is 10 by 12 ft. Compute the vertical soil pressure at each corner of the footing.

Solution. Refer to Fig. 12-1. The underlying prism of soil has a rectangular section. For this section,

$$S_x = \tfrac{1}{6}bd^2 = \tfrac{1}{6} \times 12 \times 10^2 = 200 \text{ ft}^3$$
$$S_y = \tfrac{1}{6} \times 10 \times 12^2 = 240 \text{ ft}^3$$

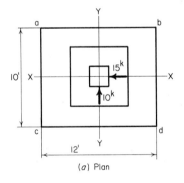

(a) Plan

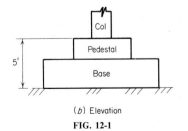

(b) Elevation

FIG. 12-1

Let f, f_x, and f_y denote the pressure caused by axial loading, bending about the x axis, and bending about the y axis, respectively.

$$f = \frac{P}{A} = \frac{250,000}{10 \times 12} = 2083 \text{ psf}$$

$$f_x = \frac{M_x}{S_x} = \frac{10,000 \times 5}{200} = 250 \text{ psf}$$

$$f_y = \frac{M_y}{S_y} = \frac{15,000 \times 5}{240} = 313 \text{ psf}$$

Then
$$f_a = 2083 + 250 + 313 = \mathbf{2646 \text{ psf}}$$
$$f_b = 2083 + 250 - 313 = \mathbf{2020 \text{ psf}}$$
$$f_c = 2083 - 250 + 313 = \mathbf{2146 \text{ psf}}$$
$$f_d = 2083 - 250 - 313 = \mathbf{1520 \text{ psf}}$$

12-2   Eccentrically Loaded Pile Groups.   When a structure is supported by vertical piles, the total load carried by the structure is transmitted to the earth as a group of concentrated vertical loads. In some instances, these loads are transmitted to the underlying soil through friction between the piles and the soil; in others, the loads are transmitted through end bearing to a deeper, sturdier stratum of soil.

A pile group is usually capped with a reinforced-concrete footing, and the rigidity of this footing serves to bind the piles to one another, causing the entire group to function as a unit. Therefore, in the absence of any special conditions, the pile group may be conceived as a structural member having a cross section that is the aggregate of the cross sections of the individual piles.

Example 12-2.   A continuous wall is founded on three rows of piles spaced 3 ft apart. The longitudinal pile spacing is 3 ft in the front and center rows and 6 ft in the rear row. The resultant of vertical loads on the wall is 20,000 plf and lies at the center row of piles. What are the pile loads in each row? Verify the results.

Solution.   Refer to Fig. 12-2. We have selected a 6-ft length as epitomizing the entire pile group. Consider the area of an individual pile to be unity. The area of the representative group is

$$A = 2 + 2 + 1 = 5$$

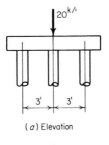

(*a*) Elevation

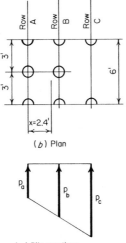

(*b*) Plan

(*c*) Pile reactions

FIG. 12-2. Eccentric load on pile group.

To locate the longitudinal centroidal axis of the group, take statical moments with respect to row A.

$$x = \frac{Q}{A} = \frac{2 \times 0 + 2 \times 3 + 1 \times 6}{5} = 2.4 \text{ ft}$$

The load on the pile group therefore has an eccentricity of

$$e = 3 - 2.4 = 0.6 \text{ ft}$$

This eccentric load can be transformed to an equivalent system consisting of a concentric load P and moment M.

$$P = 6 \times 20,000 = 120,000 \text{ lb}$$
$$M = 120,000 \times 0.6 = 72,000 \text{ ft-lb}$$

The moment of inertia of the group with respect to its centroidal axis is

$$I = 2 \times 2.4^2 + 2 \times 0.6^2 + 1 \times 3.6^2 = 25.2 \text{ ft}^2$$

Let f = pile load due to concentric load on group
f_a = pile load in row A due to moment
P_a = total pile load in row A

$$f = \frac{P}{A} = \frac{120,000}{5} = 24,000 \text{ lb per pile}$$

$$f_a = \frac{Mx}{I} = \frac{72,000 \times 2.4}{25.2} = 6860 \text{ lb per pile (uplift)}$$

Similarly, $f_b = 1710$ and $f_c = 10,290$ lb per pile

$$P_a = 24,000 - 6860 = \textbf{17,140 lb} \text{ per pile}$$
$$P_b = 24,000 + 1710 = \textbf{25,710 lb} \text{ per pile}$$
$$P_c = 24,000 + 10,290 = \textbf{34,290 lb} \text{ per pile}$$

These results can be verified by demonstrating that the aggregate load on the individual piles and the load on the pile group are equal to one another and have equal moments about any axis. Refer to the accompanying table, where moments are taken with respect to row A.

Row	Number of piles	Pile load, lb	Total load, lb	Arm, ft	Moment, ft-lb
A	2	17,140	34,280	0	0
B	2	25,710	51,420	3	154,260
C	1	34,290	34,290	6	205,740
Total			119,990		360,000

Moment of load on group = $120,000 \times 3 = 360,000$ ft-lb   OK

12-3   Criteria for Stability.   Consider that the loads acting on a structure have been resolved into their horizontal and vertical components. The horizontal components tend to cause sliding and overturning of the structure. If the structure is not anchored to its foundations, reliance must be placed on the vertical components to resist these tendencies. Therefore, stability of the structure has three requirements: the vertical soil pressure must not be excessive, the vertical components of load must be sufficiently large to offer adequate resistance to sliding, and the vertical components must be so disposed that they prevent both overturning of the structure and uplift of the structure at its heel.

12-4 Gravity Dams.

A *gravity* structure is one that carries no significant vertical load other than its own weight. Therefore, if the structure is not anchored, its weight must function as the ballast of the structure. Thus, the central problem in designing a gravity structure is to proportion it in such manner that its weight has both the required magnitude and the proper disposition.

Example 12-3. A concrete gravity dam is 20 ft high, 3 ft wide at the top, and 15 ft wide at the base. The upstream face is vertical. Compute the soil pressure at the toe and heel of the dam when the water surface is level with the top. If the coefficient of friction is 0.6, what is the factor of safety (FS) against sliding? What is the factor of safety against overturning?

Solution. Refer to Fig. 12-3a. We shall use a 1-ft length of dam as epitomizing the entire structure. The specific weights of water and reinforced concrete are 62.4 and 150 pcf, respectively. The horizontal thrust of the

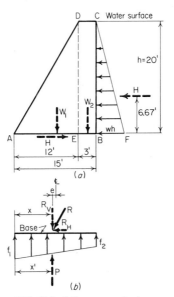

FIG. 12-3. Soil pressure under dam.

impounded fluid is represented by the pressure triangle CBF, where w denotes the specific weight of water. Let R denote the resultant of the loads on the dam, and resolve R into its components at the base, as shown in Fig. 12-3b.

Line DE in Fig. 12-3a resolves the trapezoid into two parts, and the corresponding weights are as follows:

$$W_1 = \tfrac{1}{2} \times 12 \times 20 \times 150 = 18{,}000 \text{ lb}$$
$$W_2 = 3 \times 20 \times 150 = 9000 \text{ lb}$$
$$R_V = 18{,}000 + 9000 = 27{,}000 \text{ lb}$$
$$R_H = H = \tfrac{1}{2}wh^2 = \tfrac{1}{2} \times 62.4 \times 20^2 = 12{,}480 \text{ lb}$$

Refer to Fig. 12-3b, and take moments about the toe.

$$R_V x = 18{,}000 \times 8 + 9000 \times 13.5 - 12{,}480 \times 6.67$$
$$\therefore R_V x = 144{,}000 + 121{,}500 - 83{,}200 = 182{,}300 \text{ ft-lb} \qquad (a)$$

Solving, $\qquad\qquad\qquad\qquad x = 6.75 \text{ ft}$

The prism of soil underlying this 1-ft length of dam has a rectangular cross section, 15 ft in the plane of the drawing and 1 ft normal to this plane. The eccentricity of R_V with respect to this prism is

$$e = 7.50 - 6.75 = 0.75 \text{ ft}$$

The maximum and minimum soil pressures are obtainable by Eqs. (5-19).

$$f_1 = \frac{27{,}000}{15 \times 1}\left(1 + \frac{6 \times 0.75}{15}\right)$$

or $\qquad\qquad f_1 = 1800 \times 1.3 = \textbf{2340 psf}$

and $\qquad\qquad f_2 = 1800 \times 0.7 = \textbf{1260 psf}$

$$\text{Sliding force} = R_H = 12{,}480 \text{ lb}$$
$$\text{Potential resistance to sliding} = \mu R_V = 0.6 \times 27{,}000$$
$$= 16{,}200 \text{ lb}$$
$$\text{FS against sliding} = 16{,}200/12{,}480 = \textbf{1.30}$$

From Eq. (a),

$$\text{Overturning moment} = 83{,}200 \text{ ft-lb}$$
$$\text{Stabilizing moment} = 144{,}000 + 121{,}500 = 265{,}500 \text{ ft-lb}$$
$$\text{FS against overturning} = 265{,}500/83{,}200 = \textbf{3.19}$$

The computed values of f_1 and f_2 can be verified in this manner: In Fig. 12-3b, the reaction P of the soil has the following magnitude and location:

$$P = \tfrac{1}{2} \times 15(2340 + 1260) = 27{,}000 \text{ lb} \qquad \text{OK}$$

$$x' = \frac{15}{3}\left(\frac{2340 + 2 \times 1260}{2340 + 1260}\right) = 6.75 \text{ ft} \qquad \text{OK}$$

When water flows over a dam, it is permissible to ignore the velocity effects and to assume a uniform variation of fluid pressure. For example, consider that a dam is 20 ft high and that the water surface is 21.5 ft above the bottom of the dam. The pressure diagram is a trapezoid, and the pressures at top and bottom are as follows:

$$p_{top} = 1.5 \times 62.4 = 93.6 \text{ psf}$$
$$p_{bottom} = 21.5 \times 62.4 = 1341.6 \text{ psf}$$

Refer to the discussion in Art. 5-8 concerning the "middle third" of a rectangular cross section. If the overturning moment on a dam can be increased sufficiently, a point is reached at which the heel pressure f_2 vanishes and simultaneously the action line of the resultant R intersects the base at the downstream edge of its middle third. An additional increase of this moment will cause uplift of the dam at its heel. This condition must be avoided, as it can produce detrimental effects.

12-5 Retaining Walls. The function of a retaining wall is to confine laterally a mass of earth or other granular material. With reference to Fig. 12-4, let

p = earth pressure at a distance y below point A
k = height of wall
θ = slope of surface AC
ϕ = angle of internal friction of confined material (usually taken as angle of repose)

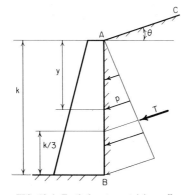

FIG. 12-4. Earth thrust on retaining wall.

According to the Rankine theory, p acts parallel to AC and its magnitude is

$$p = Cwy \tag{12-1}$$

where $$C = \cos \theta \, \frac{\cos \theta - \sqrt{\cos^2 \theta - \cos^2 \phi}}{\cos \theta + \sqrt{\cos^2 \theta - \cos^2 \phi}}$$

The total thrust T per linear foot of wall is

$$T = \tfrac{1}{2}Cwk^2$$

If θ is zero,

$$C = \frac{1 - \sin \phi}{1 + \sin \phi} \tag{12-2}$$

If nothing is stated to the contrary, we shall assume that ϕ is $33°40'$; this angle has a tangent of $\tfrac{2}{3}$.

For the special case where AC is horizontal and ϕ is $33°40'$, the preceding equations reduce to

$$p = 0.287wy \qquad \text{and} \qquad T = 0.143wk^2 \tag{12-3}$$

For the special case where both ϕ and θ are $33°40'$, we have

$$p = 0.832wy$$
and $$T = 0.416wk^2 \tag{12-4}$$

In the latter situation, however, it is expedient to resolve p and T into their horizontal and vertical components:

$$\begin{aligned} p_H &= 0.692wy & p_V &= 0.462wy \\ T_H &= 0.346wk^2 & T_V &= 0.231wk^2 \end{aligned} \tag{12-4a}$$

If the soil sustains any vertical load in addition to its own weight, the factor wy in these equations must be supplanted with the total pressure at the given point.

Example 12-4.   Determine whether the concrete retaining wall in Fig. 12-5 is safe against overturning. The weight of earth is 100 pcf.

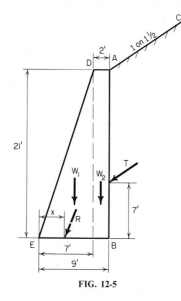

FIG. 12-5

Solution.   The action line of T intersects AB at a point 7 ft above B. Compute the moment of the loads with respect to the toe.

$$W_1 = \tfrac{1}{2} \times 7 \times 21 \times 150 = 11{,}030 \text{ lb}$$
$$W_2 = 2 \times 21 \times 150 = 6300 \text{ lb}$$
$$T_H = 0.346 \times 100 \times 21^2 = 15{,}260 \text{ lb}$$
$$T_V = 0.231 \times 100 \times 21^2 = 10{,}190 \text{ lb}$$
$$M_E = 11{,}030 \times 4.67 + 6300 \times 8 + 10{,}190 \times 9$$
$$- 15{,}260 \times 7 = 86{,}760 \text{ ft-lb}$$

Since the moment about the toe is clockwise, the wall is safe against overturning. Although the problem does not require that this be done, we shall ascertain that the resultant pierces the base within the middle third.

$$R_V = 11{,}030 + 6300 + 10{,}190 = 27{,}520 \text{ lb}$$

$$x = \frac{\Sigma M_E}{R_V} = \frac{86{,}760}{27{,}520} = 3.2 \text{ ft} > 3 \text{ ft} \qquad \text{OK}$$

Example 12-5.   Determine whether the concrete retaining wall in Fig. 12-6 is safe against overturning. Use 100 pcf for the weight of earth and 30° for the angle of repose.

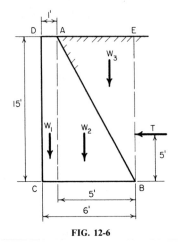

FIG. 12-6

Solution. Pass the vertical plane BE through the heel. The triangular wedge of earth AEB contributes to the stability of the wall.

$$W_1 = 2250 \text{ lb}$$
$$W_2 = 5625 \text{ lb}$$
$$W_3 = 3750 \text{ lb}$$

$$C = \frac{1 - 0.5}{1 + 0.5} = 0.333$$

$$T = \tfrac{1}{2} \times 0.333 \times 100 \times 15^2 = 3750 \text{ lb}$$
$$M_c = 2250 \times 0.5 + 5625 \times 2.67 + 3750 \times 4.33 - 3750 \times 5 = 13,630 \text{ ft-lb}$$

Since the moment is clockwise, the wall is safe against overturning.

If a wall is used merely to confine a mass of soil, the pressure exerted upon the wall is termed *active* soil pressure. If, on the other hand, a wall is used to compress the confined soil, the pressure developed in resistance to such movement is termed *passive* soil pressure. Equation (12-1) pertains to active soil pressure; the passive soil pressure is

$$p' = C'wy \tag{12-5}$$

where

$$C' = \cos \theta \, \frac{\cos \theta + \sqrt{\cos^2 \theta - \cos^2 \phi}}{\cos \theta - \sqrt{\cos^2 \theta - \cos^2 \phi}}$$

PROBLEMS

12-1. A 7- by 12-ft rectangular footing supports two columns on its longitudinal center line. The first column is centered 3 ft from the left edge and carries a load of 120 kips. The second column is centered 4 ft from the right edge and carries a load of 360 kips. What are the maximum and minimum soil pressures under the footing caused by the column loads alone?   ANS. 7860 and 3570 psf

12-2. A continuous wall is supported by the three rows of piles shown in Fig. 12-7. The longitudinal pile spacing is 3 ft in row A, 4 ft in row B, and 6 ft in row C. The total vertical load transmitted to the pile group is 18,000 plf and its resultant lies 3 ft to the right of row A. Determine the loads on the piles.

ANS. Load per pile is 26,180 lb in row A, 23,560 lb in row B, and 20,290 lb in row C

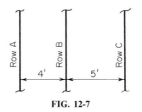

FIG. 12-7

12-3. The following information pertains to the pile group described in Problem 12-2: the pile load is 24,000 lb in row A and 26,000 lb in row B. What is the total vertical load on the pile group? Where is its resultant located?

ANS. 19,250 plf; 3.57 ft from row A

12-4. A concrete dam having a triangular profile is 15 ft high and has a base 10 ft wide. The upstream face is vertical. What is the toe pressure under the dam with the reservoir full? What is the factor of safety against overturning? Use 150 pcf as the specific weight of a concrete structure.   ANS. 2110 psf; 2.14

13

PROPERTIES OF CONCRETE—
REINFORCED-CONCRETE BEAMS

13-1 Introduction. Concrete has considerable compressive and shearing strength but only a modicum of tensile strength. Consequently, concrete is suitable as a structural material only when it is reinforced with steel in those regions where tensile stresses exist. A composite member constructed in this manner is highly economical, durable, and fire-resistant, and it serves as a good insulator against heat and sound. In this text, we shall apply the "Building Code Requirements for Reinforced Concrete" prepared by the American Concrete Institute (ACI).

When concrete is deposited in a standard test cylinder under certain prescribed conditions of temperature and moisture, the maximum compressive stress that this concrete can withstand at the expiration of 28 days is considered to be its *ultimate strength*. This value is used to identify each grade of concrete. For example, 4000-psi concrete has an ultimate strength of 4000 psi.

Steel reinforcement consists of bars of circular cross section or, in the case of light slabs, of welded-wire fabric. The bar sizes are expressed in terms of the number of eighths of an inch composing the nominal bar diameter. For example, a No. 5 bar has a nominal diameter of $\frac{5}{8}$ in. The cross-sectional areas of the bars are shown in Table 13-1. The grade of the steel is designated by its yield-point stress in ksi. For example, Grade 50 steel has a yield-point stress of 50 ksi.

TABLE 13-1   Areas of Reinforcing Bars

Bar size	Area, sq in.	Bar size	Area, sq in.
No. 3	0.11	No. 9	1.00
No. 4	0.20	No. 10	1.27
No. 5	0.31	No. 11	1.56
No. 6	0.44	No. 14	2.25
No. 7	0.60	No. 18	4.00
No. 8	0.79		

13-2   Design of Concrete Mixtures. Concrete is a mixture of inert particles, termed *aggregates*, bound together by a matrix of cement paste. The aggregates consist of two groups: fine aggregates (sand) and coarse aggregates (crushed stone, gravel, or slag). The cement paste is formed by the mixture of portland cement and water; the paste and fine aggregates constitute the mortar of the concrete.

It has been the custom to express the relative quantities of the concrete ingredients by bulk volume, in this order: cement, sand, stone. For example, a 1:2:4 mix signifies that, for every cubic foot of cement, 2 cu ft of sand and 4 cu ft of stone are to be used. The current practice is to proportion the ingredients on a gravimetric basis. Problems pertaining to the design of a concrete mixture involve an application of the following relationships:

$$1 \text{ cu ft} = 7.48 \text{ gal} \qquad 1 \text{ cu yd} = 27 \text{ cu ft}$$
$$1 \text{ cu ft of water weights } 62.4 \text{ lb}$$
$$1 \text{ gal of water weights } 8.35 \text{ lb}$$

A sack of portland cement is assumed to have a volume of 1 cu ft and to weigh 94 lb. However, only about 49 percent of this bulk, or loose, volume is occupied by solid matter, the remaining 51 percent being void. Similarly, the volumes of sand and stone are only partly filled with solid matter. The *absolute* volume corresponding to a given weight of a substance is the volume occupied solely by the solid particles. Thus the absolute volume of 94 lb of cement is 0.49 cu ft. The (absolute) specific gravity of cement is approximately 3.10, and that of the aggregates approximately 2.65.

If the amount of entrapped air is negligible, we can consider that all voids are filled when the concrete ingredients are mixed. Hence the volume of the concrete mass equals the sum of the absolute volumes of the ingredients.

Example 13-1.   A concrete mixture of $1:1\frac{1}{2}:3$ by volume with 5 gal of water per sack of cement is specified. The voids in the cement, sand, and stone are 0.51, 0.38, and 0.33, respectively. What volume of concrete per sack of cement will result?

Solution

$$\text{Volume of water} = 5/7.48 = 0.67 \text{ cu ft}$$

Ingredient	Loose volume, cu ft	×	Percent solid	=	Absolute volume, cu ft
Cement	1		49		0.49
Sand	1.5		62		0.93
Stone	3		67		2.01
Water					0.67
Volume of concrete					4.10

Example 13-2.   A concrete mixture is specified as $1:2:4$ by loose volume with 6 gal of water per sack of cement. What weight of each ingredient is required to produce 2 cu yd of concrete? Use the following data for cement, sand, and stone, respectively: percent solid, 48.5, 62, and 64; specific gravity, 3.10, 2.62, and 2.66.

Solution

$$\text{Volume of water} = 6/7.48 = 0.802 \text{ cu ft}$$

Ingredient	Loose volume, cu ft	×	Percent solid	=	Absolute volume, cu ft
Cement	1		48.5		0.485
Sand	2		62		1.240
Stone	4		64		2.560
Water					0.802
Volume of concrete per sack of cement					5.087

$$\text{Number of sacks required} = 2 \times 27/5.087 = 10.6$$

Ingredient	Final absolute volume, cu ft	×	Density, pcf	=	Weight, lb
Cement	0.485 × 10.6		3.10 × 62.4		990
Sand	1.240 × 10.6		2.62 × 62.4		2150
Stone	2.560 × 10.6		2.66 × 62.4		4500
Water	0.802 × 10.6		62.4		530

According to the solution to the preceding problem, concrete weighs 151 pcf. Reinforced concrete is generally considered to weigh 150 pcf; this total comprises 145 pcf for the concrete and 5 pcf for the steel.

The required plasticity of the concrete influences the relative proportions that the concrete ingredients are to have, and a *slump test* is performed in the laboratory to establish these relative proportions. The specimen aggregates are surface-dried before being applied in the test, and consequently the concrete mixture that evolves is based on saturated, surface-dry conditions. However, the aggregates found in the field may contain free moisture, and on the other hand aggregates that are very dry may absorb moisture during mixing. Where the aggregates will either contribute or absorb moisture, it is necessary to correct the design of the concrete mixture accordingly.

Example 13-3. A concrete mixture is to consist of the following proportions by weight: cement, 1; sand, 2.5; stone, 4.4. It will contain 5.75 gal of water per sack of cement and 3 percent entrained air. The specific gravities are as follows: cement, 3.12; sand (surface dry), 2.62; stone (surface dry), 2.69. The sand and stone contain, respectively, 5 percent and 2 percent of free moisture, based on dry weight. Specify the weight of cement, moist sand, and moist stone, and the number of gal of water, to produce 1 cu yd of concrete.

Solution. Preliminary calculations will be based on dry weights and use of 1 sack of cement.

$$\text{Volume of water} = 5.75/7.48 = 0.769 \text{ cu ft}$$

Ingredient	Weight, lb	÷ Density, pcf =	Absolute volume, cu ft
Cement	94	3.12 × 62.4	0.483
Sand	2.5 × 94 = 235	2.62 × 62.4	1.437
Stone	4.4 × 94 = 414	2.69 × 62.4	2.466
Water			0.769
Total			5.155

The total volume of concrete contains 3 percent entrained air. Then

$$\text{Volume of concrete} = 5.155/0.97 = 5.314 \text{ cu ft}$$
$$\text{Number of sacks required} = 27/5.314 = 5.08$$

The preliminary values for 1 cu yd of concrete are as follows:

$$\text{Weight of cement} = 94 \times 5.08 = 478 \text{ lb}$$
$$\text{Weight of sand} = 235 \times 5.08 = 1194 \text{ lb}$$
$$\text{Weight of stone} = 414 \times 5.08 = 2103 \text{ lb}$$
$$\text{Number of gal of water} = 5.75 \times 5.08 = 29.21$$

The corrections for free moisture are as follows:

$$\text{Moisture contributed by aggregates} = 1194 \times 0.05 + 2103 \times 0.02 = 102 \text{ lb}$$

The corrected quantities are as follows:

$$\text{Weight of cement} = 478 \text{ lb}$$
$$\text{Weight of moist sand} = 1194 \times 1.05 = 1254 \text{ lb}$$

Weight of moist stone = 2103 × 1.02 = 2145 lb
Number of gal of water = 29.21 − 102/8.35 = 16.99

Where a dry aggregate absorbs moisture during mixing, the quantity involved is usually rather small. It then suffices to increase the quantity of water to be supplied to compensate for this absorption without adjusting the quantity of the aggregate.

When dry sand is moistened, its volume swells; this phenomenon is termed *bulking*. Therefore, if the concrete ingredients are proportioned volumetrically and the sand contains moisture, the volume of sand to be supplied must be increased to allow for the bulking effect. As an illustration, assume that a specimen batch of sand is found to contract 15 percent in volume when allowed to dry and that 10 cu ft of dry sand is required. The volume of moist sand to be supplied is 10/0.85, or 11.76 cu ft.

13-3   Properties of Reinforced-Concrete Beams.   Since the tensile strength of concrete is relatively small, it is generally assumed for simplicity that the only bending stresses the concrete can sustain are those of compression. Consequently, the effective beam section is considered to comprise the reinforcing steel and the concrete on the compression side of the neutral axis, the concrete between these component areas serving merely as the ligature of the member. For further simplicity, the area of the reinforcing steel is considered to be concentrated at its centroidal axis, and as a result there is only one value of bending stress corresponding to the steel.

The spacing of steel reinforcing bars in a concrete member, such as the rectangular beam in Fig. 13-1*a* or the slab in Fig. 13-1*b*, is governed by the ACI Code. For a preliminary design, it is usually assumed that the distance from the exterior surface to the center of the first row of steel bars is $2\frac{1}{2}$ in. in a beam with web stirrups and 1 in. in a slab.

The term *depth* of beam refers to the effective rather than the overall depth of the member. It is the distance from the extreme compression fiber to the centroidal axis of the reinforcing steel. With respect to a rectangular beam, it has been found by experience that in an economically proportioned member the width-depth ratio lies between $\frac{1}{2}$ and $\frac{3}{4}$.

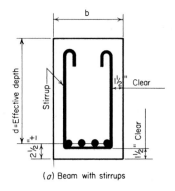

(*a*) Beam with stirrups

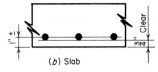

(*b*) Slab

FIG. 13-1. Spacing of reinforcing bars.

13-4   Notation.   The notational system pertaining to reinforced-concrete beams is as follows:

f'_c = ultimate strength of concrete, psi

f_c = maximum compressive stress in concrete, psi

f_s = tensile stress in steel, psi

f_y = yield-point stress of steel, psi

ϵ_c = strain of extreme compression fiber

ϵ_s = strain of steel

E_s = modulus of elasticity of steel = 29,000,000 psi

b = width of beam, in.

d = depth of beam, in.

h = thickness of slab, in.

A_s = area of tension reinforcement, sq in.

C = resultant compressive force on transverse section, lb

T = resultant tensile force on transverse section, lb

ϕ = capacity-reduction factor

ρ = tension-reinforcement ratio, A_s/bd

ω = tension-reinforcement index, $\rho f_y/f'_c$

13-5   Requirements in Beam Design.   The present ACI Code gives preference to the strength method (formerly called the ultimate-strength method) of analysis. As stated in Art. 3-10, the load carried by a member at impending failure is termed the *ultimate load*. With respect to a beam, the bending moment at the section where failure impends is termed the *ultimate moment*. The subscript *u* appended to a symbol signifies that the given quantity is being evaluated at ultimate conditions.

Figure 13-2 indicates the conditions that exist at a beam section where the ultimate moment is attained. Let c denote the distance from the neutral axis to the extreme compression fiber. To simplify the design, the assumption is made that there is a uniform stress f_c in the concrete extending across a depth a, and that

$$f_c = 0.85f'_c \qquad \text{and} \qquad a = \beta_1 c$$

where β_1 has the value stipulated in the ACI Code.

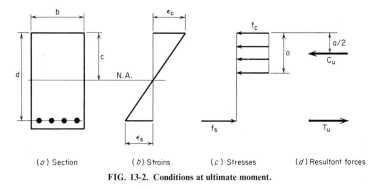

(a) Section   (b) Strains   (c) Stresses   (d) Resultant forces

FIG. 13-2. Conditions at ultimate moment.

A reinforced-concrete beam has three potential modes of flexural failure: (1) crushing of the concrete, which is assumed to occur when ϵ_c reaches the value 0.003; (2) yielding of the steel, which begins when f_s reaches the value f_y; and (3) the simultaneous crushing of the concrete and yielding of the steel. A beam that tends to fail by the third mode is said to be *balanced*. If the value of ρ exceeds that corresponding to balanced conditions (i.e., if the reinforcement is excessive), the beam tends to fail by crushing of the concrete. On the other hand, if the value of ρ is less than that corresponding to balanced conditions, the beam tends to fail by yielding of the steel.

Failure of the beam by the first mode would occur precipitously and without warning, whereas failure by the second mode would occur gradually, offering visible evidence of progressive failure. Therefore, to ensure that yielding of the steel would precede failure of the concrete, the ACI Code limits the amount of reinforcement that may be supplied to 0.75 times that corresponding to balanced conditions. In addition, the Code establishes a minimum value of reinforcement, setting $\rho_{min} = 200/f_y$ where f_y is in psi. Then

$$A_{s,min} = 200bd/f_y$$

To allow for imperfections of material, defects in workmanship, etc., the Code introduces the capacity-reduction factor ϕ. This has the value 0.90 with respect to flexure.

13-6   Rectangular Beams.   The basic equations for the strength design of a rectangular beam reinforced solely in tension are recorded below. The subscript b appended to a symbol signifies that the given quantity is being evaluated at balanced conditions. In Eqs. (13-8) to (13-12), inclusive, f_y is in psi. Equation (13-8) is obtained by setting the strain of the steel at incipient yielding equal to f_y/E_s.

$$C_u = 0.85abf'_c \qquad T_u = A_s f_y \tag{13-1}$$

$$\omega = \frac{A_s\, f_y}{bd\, f'_c} \tag{13-2}$$

$$a = 1.18\omega d \qquad c = \frac{1.18\omega d}{\beta_1} \tag{13-3}$$

$$M_u = \phi A_s f_y \left(d - \frac{a}{2} \right) \tag{13-4}$$

$$M_u = \phi A_s f_y d(1 - 0.59\omega) \tag{13-5}$$

$$M_u = \phi bd^2 f'_c \omega(1 - 0.59\omega) \tag{13-6}$$

$$A_s = \frac{bdf_c - \sqrt{(bdf_c)^2 - 2bf_c M_u/\phi}}{f_y} \tag{13-7}$$

$$c_b = \frac{87,000}{87,000 + f_y}\, d \tag{13-8}$$

$$\rho_b = 0.85\beta_1 \frac{f'_c}{f_y} \frac{87,000}{87,000 + f_y} \tag{13-9}$$

$$\omega_b = 0.85\beta_1 \frac{87,000}{87,000 + f_y} \tag{13-10}$$

Applying the limitation imposed on the amount of reinforcement, we obtain

$$\rho_{max} = 0.75\rho_b = 0.6375\beta_1 \frac{f'_c}{f_y} \frac{87,000}{87,000 + f_y} \qquad (13\text{-}11)$$

$$\omega_{max} = 0.6375\beta_1 \frac{87,000}{87,000 + f_y} \qquad (13\text{-}12)$$

Figure 13-3 indicates the relationship between M_u and A_s for a beam of given size. As A_s increases, the internal forces C_u and T_u increase proportionately, but M_u increases by a smaller proportion because the action line of C_u is depressed. The M_u-A_s diagram is parabolic, but its curvature is slight. Comparing the coordinates of two points P_a and P_b, we obtain the following relationship, in which subscripts a and b refer to P_a and P_b, respectively:

$$\frac{M_{ua}}{A_{sa}} > \frac{M_{ub}}{A_{sb}} \qquad \text{if } A_{sa} < A_{sb} \qquad (13\text{-}13)$$

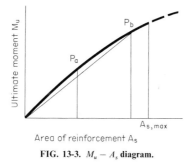

FIG. 13-3. $M_u - A_s$ diagram.

Example 13-4. A rectangular beam having a width of 12 in. and effective depth of 19.5 in. is reinforced with steel bars having an area of 5.37 sq in. The beam is made of 2500-psi concrete and the steel is of Grade 40. Compute the ultimate moment that this beam can resist (*a*) without referring to any design tables and without applying the basic equations except those that are readily apparent and (*b*) by applying the basic equations.

Solution

$$A_{s,min} = 200 \times 12 \times 19.5/40,000 = 1.17 \text{ sq in.} \qquad \text{OK}$$

PART *a*: Refer to Fig. 13-2. We shall first determine whether the area of reinforcement is below the upper limit imposed by the ACI Code. The following calculations pertain to a balanced beam.

$$\varepsilon_s = \frac{f_y}{E_s} = \frac{40{,}000}{29{,}000{,}000} = 0.00138$$

$$\frac{c}{d} = \frac{\varepsilon_c}{\varepsilon_c + \varepsilon_s} = \frac{0.003}{0.003 + 0.00138} = 0.685$$

$$c = 0.685 \times 19.5 = 13.36 \text{ in.}$$

For 2500-psi concrete, $\beta_1 = 0.85$. Then

$$a = \beta_1 c = 0.85 \times 13.36 = 11.36 \text{ in.}$$
$$f_c = 0.85 f_c' = 0.85 \times 2500 = 2125 \text{ psi}$$
$$T_u = C_u = abf_c = 11.36 \times 12 \times 2125 = 290{,}000 \text{ lb}$$

$$A_s = \frac{T_u}{f_y} = \frac{290{,}000}{40{,}000} = 7.25 \text{ sq in.}$$

With respect to the true beam,

$$A_{s,\text{max}} = 0.75 \times 7.25 = 5.44 > 5.37 \text{ sq in.}$$

and the member is therefore satisfactory in this respect.

The calculations for ultimate moment are as follows:

$$T_u = A_s f_y = 5.37 \times 40{,}000 = 215{,}000 \text{ lb}$$
$$C_u = abf_c = a \times 12 \times 2125 = 215{,}000 \text{ lb} \qquad \therefore a = 8.43 \text{ in.}$$

$$M_u = \phi T_u \left(d - \frac{a}{2} \right) = 0.90 \times 215{,}000 \left(19.5 - \frac{8.43}{2} \right)$$

$$= \mathbf{2{,}960{,}000 \text{ in.-lb}}$$

PART *b*: Applying Eqs. (13-2) and (13-12), we have

$$\omega = \frac{5.37}{12 \times 19.5} \frac{40}{2.5} = 0.367$$

$$\omega_{\text{max}} = 0.6375 \times 0.85 \times \tfrac{87}{127} = 0.371 \qquad\qquad \text{OK}$$

Equation (13-5) yields

$$M_u = 0.90 \times 5.37 \times 40{,}000 \times 19.5(1 - 0.59 \times 0.367)$$
$$= 2{,}960{,}000 \text{ in.-lb}$$

Example 13-5.   A beam on a simple span of 20 ft is to carry a uniformly distributed live load of 1770 plf and dead load of 500 plf, which includes the estimated weight of beam. Architectural details restrict the beam width to 12 in. and require that the depth be made as small as possible. Design the section, using 3000-psi concrete and Grade 40 steel.

Solution. The beam depth is minimized by providing the maximum amount of reinforcement permitted by the Code. For 3000-psi concrete, $\beta_1 = 0.85$, and from Example 13-4 we have $\omega_{max} = 0.371$. Applying the load factors prescribed by the Code, we obtain the following value of uniformly distributed ultimate load:

$$w_u = 1.4 \times 500 + 1.7 \times 1770 = 3710 \text{ plf}$$
$$M_u = \tfrac{1}{8} \times 3710 \times 20^2 \times 12 = 2{,}230{,}000 \text{ in.-lb}$$

By Eq. (13-6),

$$d^2 = \frac{M_u}{\phi b f'_c \omega (1 - 0.59\omega)}$$

$$= \frac{2{,}230{,}000}{0.90 \times 12 \times 3000 \times 0.371 \times 0.781} \qquad d = 15.4 \text{ in.}$$

Set $d = 15.5$ in. to make the overall depth an integral number of inches. The corresponding reduction in the value of ω is negligible, and Eq. (13-2) may now be solved for the area of reinforcement.

$$A_s = \omega b d \frac{f'_c}{f_y} = 0.371 \times 12 \times 15.5 \times \tfrac{3}{40} = 5.18 \text{ sq in.}$$

Use four No. 9 and two No. 7 bars, for which $A_s = 5.20$ sq in.

The beam is presumed to have adequate stiffness if the ratio of span to overall depth does not exceed the limiting value established by the ACI Code. For a simply supported beam, this value is 16. In the present case, the ratio is at most

$$20 \times 12/18 < 16$$

Therefore, it is not necessary to investigate the deflection of the beam under service loads.

To provide a compact spacing, the ACI Code allows parallel reinforcing bars of No. 11 or smaller size to be bundled, the number of bars in a bundle being restricted to four. However, bundled bars are subject to special provisions, and we shall therefore assume in this text that bars will not be bundled. To maintain the minimum distance between bars, it is necessary to place the bars we have selected in two rows. Taking 3.5 in. as the approximate distance from the outer edge to the centroidal axis of the steel area, we obtain an overall depth of $15.5 + 3.5 = 19$ in.

In summary, the design is as follows:

Beam size: 12 by 19 in.
Reinforcement: Four No. 9 and two No. 7 bars.

Example 13-6. A beam is to sustain an ultimate moment of 305 ft-kips. Determine the width, effective depth, and area of reinforcement, using 3000-psi concrete and Grade 60 steel.

Solution. The design will be undertaken by using the maximum value of ω. By Eq. (13-12),

$$\omega_{max} = 0.6375 \times 0.85 \times \tfrac{87}{147} = 0.321$$

By Eq. (13-6),

$$bd^2 = \frac{M_u}{\phi f'_c \omega (1 - 0.59\omega)}$$

$$= \frac{305 \times 12}{0.90 \times 3 \times 0.321 \times 0.811} = 5207 \text{ in.}^3$$

This requirement is satisfied by $b = 11.5$ in. and $d = 21.5$ in., giving $bd^2 = 5316$ in.^{3}

Since the true value of ω will differ appreciably from that of ω_{max}, the area of reinforcement must be found by applying Eq. (13-7).

$$f_c = 0.85 \times 3 = 2.55 \text{ ksi}$$
$$bd f_c = 11.5 \times 21.5 \times 2.55 = 630.5 \text{ kips}$$

$$\frac{2bf_c M_u}{\phi} = \frac{2 \times 11.5 \times 2.55 \times 305 \times 12}{0.90} = 238{,}500 \text{ kips}^2$$

$$A_s = \frac{630.5 - \sqrt{630.5^2 - 238{,}500}}{60} = 3.86 \text{ sq in.}$$

The true value of ω is

$$\omega = \frac{3.86}{11.5 \times 21.5} \frac{60}{3} = 0.312$$

13-7   T Beams. A reinforced-concrete girder is generally constructed integrally with the slab that it supports. It is therefore reasonable to consider that the beam itself and a portion of the tributary slab constitute a single structural entity in the form of a T-shaped beam. The upper and lower parts of the T are referred to as the flange and web (or stem), respectively. The effective width of flange is determined by applying the provisions of the ACI Code. Isolated T beams are also used when the depth is limited by architectural features.

If the compression block lies wholly within the flange, the design of the T beam is identical with that of a rectangular beam in all respects except that the minimum area of reinforcement is computed by applying the width of web. The notational system for a T beam, most of which is illustrated in Fig. 13-4, is as follows:

b = width of flange
b_w = width of web
t = thickness of flange
a = total depth of compression block
m = depth of compression block in web = $a - t$
C_{uf} = resultant compressive force in flange
C_{uw} = resultant compressive force in web
M_{uf} = bending-moment capacity associated with C_{uf}
M_{uw} = bending-moment capacity associated with C_{uw}

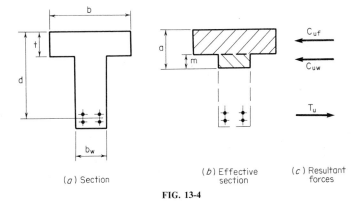

(a) Section   (b) Effective section   (c) Resultant forces

FIG. 13-4

Consider the tensile force T_u to be resolved into two components having the magnitudes C_{uf} and C_{uw} and both located at the centroidal axis of the steel. The resultant force system on the section now consists of two couples having the following moment capacities:

$$M_{uf} = \phi C_{uf}(d - \tfrac{1}{2}t) = \phi btf_c(d - \tfrac{1}{2}t) \qquad (13\text{-}14a)$$
$$M_{uw} = \phi C_{uw}(d - t - \tfrac{1}{2}m) = \phi b_w m f_c(d - t - \tfrac{1}{2}m) \quad (13\text{-}14b)$$

and $$M_u = M_{uf} + M_{uw}$$

Equation (13-14b) can be transformed to the following:

$$m^2 - 2m(d - t) + \frac{2M_{uw}}{\phi b_w f_c} = 0 \qquad (13\text{-}15)$$

This equation offers a means of finding m when M_{uw} is known.

Example 13-7. The T beam in Fig. 13-5 is made of 3000-psi concrete and is reinforced with Grade 40 steel. Determine the ultimate moment in in.-lb that this member can resist.

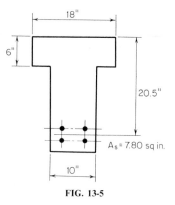

18"

6"

20.5"

$A_s = 7.80$ sq in.

10"

FIG. 13-5

Solution. We shall first determine whether the area of reinforcement falls within the allowable range.

$$A_{s,min} = 200 b_w d/f_y = 200 \times 10 \times 20.5/40{,}000 = 1.03 \text{ sq in.}$$

Consider a balanced beam having the given dimensions. By Eq. (13-8),

$$c_b = \tfrac{87}{127} \times 20.5 = 14.04 \text{ in.}$$

$$a_b = 0.85 \times 14.04 = 11.93 \text{ in.} \qquad m = 11.93 - 6 = 5.93 \text{ in.}$$

$$f_c = 2550 \text{ psi}$$

$$C_{ub} = 2550(6 \times 18 + 5.93 \times 10) = 426{,}600 \text{ lb}$$

$$A_{sb} = \frac{T_{ub}}{f_y} = \frac{426{,}600}{40{,}000} = 10.67 \text{ sq in.}$$

With respect to the true beam,

$$A_{s,\text{max}} = 0.75 \times 10.67 = 8.00 \text{ sq in.}$$

The area of reinforcement therefore satisfies the imposed limits.

$$C_u = T_u = A_s f_y = 7.80 \times 40,000 = 312,000 \text{ lb}$$
$$C_{uf} = 6 \times 18 \times 2550 = 275,400 \text{ lb}$$
$$C_{uw} = 312,000 - 275,400 = 36,600 \text{ lb}$$

$$m = \frac{36,600}{2550 \times 10} = 1.44 \text{ in.}$$

By Eqs. (13-14),

$$M_u = 0.90[275,400(20.5 - 3) + 36,600(20.5 - 6 - 0.72)]$$
$$= \textbf{4,790,000 in.-lb}$$

13-8 Doubly Reinforced Beams. Where the architectural details limit both the depth and width of a concrete beam, it may be necessary to incorporate steel reinforcement in the compression region as well as the tension region of the member, as illustrated in Fig. 13-6. For simplicity, we shall disregard the reduction of concrete area caused by the presence of the compression reinforcement. The notational system is as follows:

A_s = area of tension reinforcement
A'_s = area of compression reinforcement
d' = distance from compression face of concrete to centroid of compression steel
f_s = stress in tension steel
f'_s = stress in compression steel
ϵ'_s = strain of compression reinforcement
ϵ_y = strain of steel at incipient yielding = f_y/E_s
ρ = A_s/bd
ω = $\rho f_y/f'_c$
M_u = ultimate moment to be resisted by member
M_{u1} = ultimate-moment capacity of member if reinforced solely in tension
M_{u2} = increase in ultimate-moment capacity accruing from use of compression reinforcement

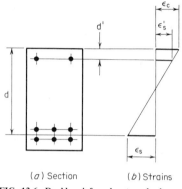

(*a*) Section   (*b*) Strains

FIG. 13-6. Doubly reinforced rectangular beam.

The tension reinforcement may be resolved into two parts. The first part, denoted by A_{s1}, acts in combination with the concrete to develop the moment M_{u1}. The second part, denoted by A_{s2}, acts in combination with the compression reinforcement to develop the moment M_{u2}. The 1963 ACI Code contained the following provisions with respect to a doubly reinforced rectangular beam:

1. It allowed us to set $f'_s = f_y$ if the beam dimensions are such that $\epsilon'_s = \epsilon_y$ at ultimate moment under balanced conditions.

2. It required that A_{s1} be limited to 0.75 times the value corresponding to balanced conditions.

The present ACI Code makes no special provision for doubly reinforced beams, and it simply subjects the total tension-reinforcement area A_s to the specified limit.

Where we may set $f'_s = f_y$, the following equations apply:

$$(A_s - A'_s)f_y = 0.85abf'_c \qquad (13\text{-}16)$$
$$M_{u2} = \phi A'_s f_y(d - d') \qquad (13\text{-}17)$$

$$M_u = \phi f_y \left[(A_s - A'_s)\left(d - \frac{a}{2}\right) + A'_s(d - d') \right] \qquad (13\text{-}18)$$

Example 13-8.   A beam that is to resist an ultimate moment of 690 ft-kips is restricted to a 14-in. width and 24-in. total depth. Using 5000-psi concrete and Grade 50 steel, determine the area of reinforcement in accordance with the 1963 ACI Code. Verify the design.

Solution

$$M = 690,000 \times 12 = 8,280,000 \text{ in.-lb}$$

We shall calculate the value of M_{u1} to ascertain whether tension reinforcement alone will suffice. The value of β_1 for this grade of concrete is 0.80. Since two rows of tension bars are undoubtedly required,

$$d = 24 - 3.5 = 20.5 \text{ in.}$$

Applying Eq. (13-10), we obtain the following values for a singly reinforced beam:

$$\omega_b = 0.85 \times 0.80 \times \tfrac{87}{137} = 0.432$$
$$\omega_{max} = 0.75 \times 0.432 = 0.324$$

By Eq. (13-6),

$$M_{u1} = 0.90 \times 14 \times 20.5^2 \times 5000 \times 0.324 \times 0.809$$
$$= 6,940,000 \text{ in.-lb}$$

The member therefore requires compression reinforcement, and

$$M_{u2} = 8,280,000 - 6,940,000 = 1,340,000 \text{ in.-lb}$$

We must investigate the strain in the compression reinforcement at ultimate moment under balanced conditions. Equation (13-3) yields

$$c_b = \frac{1.18\omega_b d}{\beta_1} = \frac{1.18 \times 0.432 \times 20.5}{0.80} = 13.1 \text{ in.}$$

Refer to Fig. 13-6b, and set $d' = 2.5$ in.

$$d - c_b = 20.5 - 13.1 = 7.4 \text{ in.} \qquad c - d' = 10.6 \text{ in.}$$

Since the strain of the tension steel is ϵ_y at ultimate moment and the compression steel is more remote from the neutral axis than is the tension steel at balanced conditions, it follows that $\epsilon'_s > \epsilon_y$, and we therefore set $f'_s = f_y$ in the true beam. By Eq. (13-17),

$$A'_s = \frac{1,340,000}{0.90 \times 50,000 \times 18.0} = 1.65 \text{ sq in.}$$

Equation (13-2) becomes

$$A_s - A'_s = \omega_{max} bd \frac{f'_c}{f_y}$$
$$= 0.324 \times 14 \times 20.5 \times \tfrac{5}{50} = 9.30 \text{ sq in.}$$
$$A_s = 9.30 + 1.65 = \textbf{10.95 sq. in.}$$

We shall verify the design. Equation (13-16) yields the following:

$$a = \frac{(A_s - A'_s)f_y}{0.85f'_c b} = \frac{9.30 \times 50,000}{0.85 \times 5000 \times 14} = 7.82 \text{ in.}$$

By Eq. (13-18), the moment capacity is

$$M_u = 0.90 \times 50,000(9.30 \times 16.59 + 1.65 \times 18)$$
$$= 8,280,000 \text{ in.-lb} \qquad \text{OK}$$

Example 13-9. With reference to Example 13-8, what modification in the design is required to conform to the present ACI Code?

Solution. If the restriction on the total tension reinforcement is interpreted literally, it requires that both A_{s1} and A_{s2} be limited to 0.75 times the value corresponding to balanced conditions. The limit on A_{s1} has been included in the present design, but the limit on A_{s2} has not. Technically, this limit can be met without impairing the capacity of the member simply by increasing A'_s by one-third its present value and retaining the present value of A_s. This increase in the area of compression reinforcement is intended solely to comply with the letter of the present Code, but it is otherwise devoid of logic.

13-9   Shearing Stress, Bond Stress, and Development Length.   The analysis of beams in Art. 5-6 demonstrated that horizontal shearing forces are present throughout the depth of a beam. It will be advantageous at this point to review the subject of horizontal shearing forces with reference to the particular case of a reinforced-concrete beam.

Figure 13-7a is a free-body diagram of a segment of a rectangular beam bounded by cross sections *a* and *b*, which are an infinitesimal distance *dx* apart. External load is assumed to be absent, and the

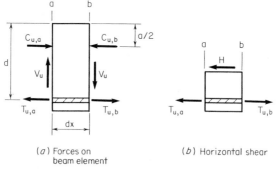

(*a*) Forces on   (*b*) Horizontal shear
beam element

FIG. 13-7

ultimate vertical shear at the boundary sections is denoted by V_u. If the bending moment at b exceeds that at a, the resultant compressive and tensile forces at b exceed those at a. The horizontal shearing force H has its maximum value between the lower edge of the compression zone and the reinforcing steel, being equal to $T_{u,b} - T_{u,a}$ across this depth, as shown in Fig. 13-7b.

To simplify the analysis of shearing stresses in the concrete, the ACI Code provides for the calculation of a *nominal* shearing stress v_u as given by the equation

$$v_u = \frac{V_u}{\phi bd} \tag{13-19}$$

where $\phi = 0.85$. For an I or T beam, b is replaced with the width of stem.

The integrity of a reinforced-concrete beam is maintained by bond between the steel and concrete, which prevents the longitudinal movement of one material relative to the other and thereby constrains the two materials to deform conjunctively. Earlier design codes required that the bond stress at every point along the span be less than a specified allowable stress. However, the present ACI Code disregards local bond stress and recognizes that relative movement of the two materials is precluded if the reinforcing bars are embedded in the concrete for a sufficient distance at their ends to develop their strength by bond. The distance across which the capacity of a bar is developed by bond is termed its *development length*. Let

  l_d = development length, in.
  A_b = cross-sectional area, sq in.
  d_b = nominal diameter of bar, in.
The equation for basic development length is

$$l_d = 0.04 A_b f_y / \sqrt{f_c'} \qquad \text{or} \qquad l_d = 0.0004 d_b f_y \tag{13-20}$$

The greater value applies. The basic development length is multiplied by an appropriate factor to suit the particular conditions. For example, with respect to reinforcement that lies at the top of a member and has more than 12 in. of concrete below it, the modification factor is 1.4.

Example 13-10. What is the basic development length of a No. 7 bar of Grade 40 steel embedded in 3000-psi concrete?

Solution. Applying both forms of Eq. (13-20), we obtain

$$l_d = 0.04 \times 0.60 \times 40{,}000/\sqrt{3000} = 17.5 \text{ in.}$$
$$l_d = 0.0004 \times 0.875 \times 40{,}000 = 14 \text{ in.}$$

The basic development length is **17.5 in.**

Where there is insufficient space to provide the full development length of a reinforcing bar, recourse must be had to use of some mechanical anchoring device or to hooking the bar at its ends. The ACI Code establishes the value of a standard form of hook. At any rate, the capacity of a bar is limited to the axial force that it acquires through anchorage in the concrete.

Consider that a simply supported beam of relatively short span carries a uniformly distributed load, and assume tentatively that all reinforcing bars terminate at the center of support. In the elastic range, the stress in the bars is directly proportional to the bending moment, and the parabolic arc *a* in Fig. 13-8 represents the manner in which the axial force induced in the reinforcement by flexure varies across the half-span. Strength is imparted to the reinforcement by bond at a uniform rate across the end distance l_d, as represented by the straight line *b*. This diagram reveals that the flexural requirements exceed the strength of the reinforcing bars over a considerable part of the span, notwithstanding the fact that l_d is less than half the span.

To preclude a condition of this type, it is necessary to use bars of relatively large size so that the slope of line *b* in Fig. 13-8 is at least equal to the slope of arc *a* at the support. Bending stress varies with

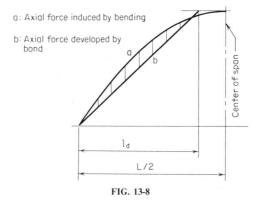

a: Axial force induced by bending

b: Axial force developed by bond

l_d

L/2

Center of span

FIG. 13-8

bending moment, and bending moment varies with shear in accordance with the equation $dM/dx = V$. Therefore, the slope of arc a at the support is directly proportional to the shear at that section. On the basis of these considerations, the ACI Code establishes the following requirement with respect to the tension bars in a simply supported beam:

$$l_d \leqslant \frac{M_t}{V_u} + l_a \qquad (13\text{-}21)$$

where M_t = flexural capacity of the bars

V_u = vertical shear at the support

l_a = sum of embedment length beyond center of support and equivalent embedment length secured by standard hook or mechanical device

If only part of the reinforcement extends to the support, M_t is taken as the moment capacity of those bars alone. An analogous requirement applies to the bars that terminate within the span, but they terminate at a section of lower shear value.

If a beam is partly restrained against rotation at the supports, the bending moment undergoes a change in algebraic sign along the span. The basic equation of the elastic curve of a beam, $d\theta/dx = -M/EI$, reveals that a section of zero moment corresponds to a point of inflection in the elastic curve. The positive reinforcement that extends to a point of inflection is subject to the same problem concerning development length as is the reinforcement at the support in a simple beam. For a beam having both positive and negative bending, the symbols in Eq. (13-21) acquire these meanings:

V_u = vertical shear at point of inflection

l_a = effective depth of beam or twelve bar diameters, whichever is greater

13-10 Reinforcement for Diagonal Tension. The principles developed in Art. 3-3 lead to the conclusion that at every point in a reinforced-concrete beam between the compression zone and the longitudinal reinforcement there exists a tensile stress that is numerically equal to the shearing stress at that point and inclined at $45°$ with the longitudinal axis. It is assumed that this is the maximum diagonal tensile stress at the section. If this normal stress is excessive, steel reinforcement must be supplied to prevent failure by diagonal tension. This reinforcement usually consists of vertical stirrups looped about the longitudinal reinforcement and extended into the compression zone for anchorage, as shown in Fig. 13-1a. Following standard practice, we shall refer to the diagonal tensile stress as a "shearing stress," although this term is a misnomer. Let

v_c = nominal shearing stress resisted by concrete
v_s = nominal shearing stress resisted by stirrup
A_v = total cross-sectional area of stirrup
F = force carried by vertical stirrup
s = center-to-center spacing of stirrups

The ACI Code provides two alternative methods of computing v_c. We shall apply the simpler method, which relates this stress solely to the ultimate strength of the concrete. The relationship is

$$v_c = 2\sqrt{f'_c} \qquad (13\text{-}22)$$

If the stirrup can be fully developed, the force in the stirrup is $A_v f_y$. If the shearing stress is constant along a distance s, the force exerted on a vertical stirrup at the center of that region is $sv_s b$. Then

$$F = A_v f_y = sv_s b \qquad (13\text{-}23)$$

The Code places a lower limit on the beam size by placing the following limit on v_s:

$$v_{s,\text{allow}} = 8\sqrt{f'_c} \qquad (13\text{-}24)$$

Where $v_u \geqslant 0.5v_c$, the Code requires that a specified minimum amount of reinforcement be supplied. For vertical stirrups, this requirement is tantamount to designing for a shearing stress v_s of 50 psi.

Let A denote the section at face of support and B the section at a distance d from A. For the region AB, the critical shearing stress is taken as the value of v_u at B.

Example 13-11. A beam of 15-in. width and 22.5-in. effective depth carries a uniformly distributed ultimate load of 10.2 klf. The beam is simply supported, and the clear distance between supports is 18 ft. Using $f'_c = 3000$ psi and $f_y = 40,000$ psi, design shear reinforcement in the form of vertical U stirrups.

Solution.   By Eqs. (13-22) and (13-24),

$$v_c = 2\sqrt{3000} = 110 \text{ psi} \qquad v_{s,\text{allow}} = 8\sqrt{3000} = 438 \text{ psi}$$

The total load on the clear span is

$$W_u = 10,200 \times 18 = 183,600 \text{ lb}$$

By Eq. (13-19), the shearing stress at face of support is

$$v_u = \frac{\frac{1}{2} \times 183{,}600}{0.85 \times 15 \times 22.5} = 320 \text{ psi}$$

Refer to Fig. 13-9.

$$\text{Slope of diagram} = -320/108 = -2.96 \text{ psi per in.}$$

At distance d from face of support,

$$v_u = 320 - 22.5 \times 2.96 = 253 \text{ psi}$$
$$v_s = v_u - v_c = 253 - 110 = 143 < 438 \text{ psi}$$

The beam size is therefore adequate with respect to shear.

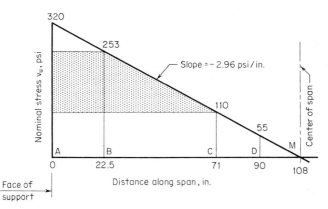

FIG. 13-9. Shearing-stress diagram.

The section C at which $v_u = v_c = 110$ psi has this location:

$$AC = \frac{320 - 110}{2.96} = 71 \text{ in.}$$

The section D at which $v_u = 0.5v_c = 55$ psi has this location:

$$AD = \frac{320 - 55}{2.96} = 90 \text{ in.}$$

To select the stirrup size, we shall equate the spacing near the support to the minimum practical value, which is generally considered to be 4 in. Equation (13-23) yields

$$A_v = \frac{sv_s b}{f_y} = \frac{4 \times 143 \times 15}{40{,}000} = 0.215 \text{ sq in.}$$

Since each stirrup is bent into the form of a U, the total cross-sectional area is twice that of a straight bar. Try No. 3 stirrups, for which

$$A_v = 2 \times 0.11 = 0.22 \text{ sq in.}$$

The basic development length as given by Eq. (13-20) is

$$l_d = 0.0004 \times \tfrac{3}{8} \times 40,000 = 6 \text{ in.}$$

However, a stirrup is subject to the additional requirement that the length of embedment must be at least 24 bar diameters, or 9 in. in the present instance. The anchorage is considered to begin at mid-depth of the member. Allowing 1.5 in. of cover, the available length of embedment is

$$\tfrac{1}{2} \times 22.5 - 1.5 = 9.75 \text{ in.} \qquad \text{OK}$$

∴ Use No. 3 stirrup. (For a shallower beam, anchorage may be achieved by hooking the bar in the standard manner and providing half the basic development length. If complete development is impossible to obtain, the stress in the stirrup must be reduced proportionately.)

The upper limits for stirrup spacing imposed by the Code are as follows:

$$s_{max} = d/4 \qquad \text{if} \qquad v_s > 4\sqrt{f'_c}$$
$$s_{max} = d/2 \qquad \text{if} \qquad v_s \leqslant 4\sqrt{f'_c}$$

In the present instance,

$$4\sqrt{f'_c} = 219 > 143 \text{ psi}$$

Therefore, the $d/4$ limit does not apply.

The required stirrup spacing as given by Eq. (13-23) is

$$s = \frac{A_v f_y}{v_s b} = \frac{0.22 \times 40,000}{15 v_s} = \frac{586.7}{v_s} \qquad (a)$$

Applying the prescribed value $v_s = 50$ psi for the region CD where minimum reinforcement is required, we obtain

$$s = \frac{586.7}{50} = 11.7 \text{ in.}$$

A satisfactory disposition of stirrups can be formulated most readily by constructing a stirrup-spacing diagram, as shown in Fig. 13-10. This diagram consists of vertical lines, each line representing a stirrup. The horizontal location of the line represents the location of the corresponding stirrup along the span, and the length of the line represents the distance between the corresponding stirrup and its successor.

Taking values of v_s along the span and computing the corresponding values of s as given by Eq. (a), we construct the diagram of maximum stirrup spacing in Fig. 13-10. From A to C, the upper limit for s is $d/2 = 11.25$ in.; from C

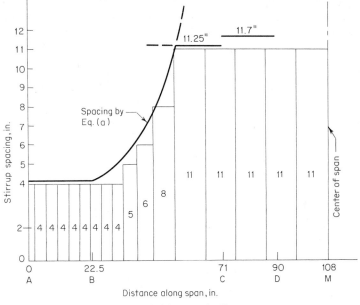

FIG. 13-10. Stirrup-spacing diagram.

to D, it is 11.7 in. Having plotted the limits for s, we then devise a stirrup spacing that conforms to those limits. The spacing we have established, which requires 33 stirrups for the entire span, is shown in Fig. 13-10. The numbers in this drawing between the vertical lines are the distances between the corresponding stirrups. Although reinforcement is not needed between D and M, stirrups have nevertheless been supplied throughout the span for simplicity. The arrangement of stirrups is also recorded in Table 13-2.

TABLE 13-2 Disposition of Stirrups

Quantity	Spacing, in.	Total, in.	Cumulative total, in.
1	2	2	2
8	4	32	34
1	5	5	39
1	6	6	45
1	8	8	53
5	11	55	108

13-11   Continuous Beams.   A concrete structure that is built monolithically is composed of rigid frames rather than isolated beams and columns, and therefore many reinforced-concrete beams are continuous over several spans. Because a precise evaluation of the maximum potential positive and negative bending moments in each span is laborious and time-consuming, the ACI Code permits the use of a set of standard moment equations if the span and load conditions satisfy the imposed restrictions. The notational system for the span is as follows:

L = center-to-center spacing of supports

L_n = clear distance between supports

Example 13-12.   A floor slab that is continuous over several spans carries a live load of 225 psf and a dead load of 50 psf, exclusive of its own weight. The clear spans are 14 ft. Design the interior span, using $f'_c = 3000$ psi and $f_y = 50,000$ psi.

Solution.   Refer to Fig. 13-11. A slab is designed by considering a 1-ft strip as an individual beam, thus making $b = 12$ in. With the beam weight included, the dead load will exceed one-third of the live load, and the standard moment equations are applicable. In compliance with the ACI Code, the slab will be designed for a maximum positive bending moment of $\frac{1}{16}w_u L_n{}^2$ at mid-span, a maximum negative bending moment of $\frac{1}{11}w_u L_n{}^2$ at the face of support, and a maximum shear of $\frac{1}{2}w_u L_n$. The steel reinforcement will consist of both straight bars and trussed bars, the latter acting as positive reinforcement in the central region and as negative reinforcement in the end regions.

Concrete tends to contract as it hardens, and its volume also tends to fluctuate with changes in temperature. Since the concrete is not free to contract, tensile stresses are engendered in all directions, and the slab must therefore be reinforced transversely as well as longitudinally.

To prevent excessive deflection, the Code imposes a lower limit on the slab thickness h. Assume that $L = 15$ ft. Then

$$h_{min} = \frac{L}{28} = \frac{15 \times 12}{28} = 6.4 \text{ in.}$$

We shall therefore try a 7-in. slab, for which $d = 7 - 1 = 6$ in. Since a slab is not reinforced against diagonal tension, it is necessary to investigate its capacity in shear as well as in flexure.

$$\text{Beam weight} = \frac{7}{12} \times 150 = 88 \text{ plf}$$
$$w_u = 1.4(50 + 88) + 1.7 \times 225 = 576 \text{ plf}$$

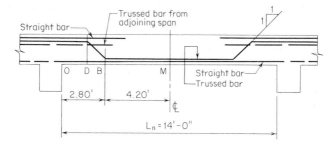

(a) Arrangement of reinforcing bars

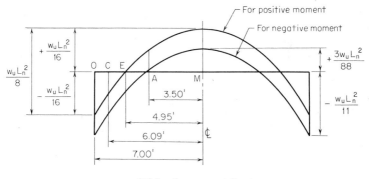

(b) Bending-moment diagrams

FIG. 13-11

For simplicity, consider the shear at face of support.

$$V_u = \tfrac{1}{2} \times 576 \times 14 = 4030 \text{ lb}$$

$$v_u = \frac{4030}{0.85 \times 12 \times 6} = 66 \text{ psi} \qquad v_c = 2\sqrt{3000} = 110 \text{ psi} \qquad \text{OK}$$

$$M_{u,\text{neg}} = \tfrac{1}{11} \times 576 \times 14^2 \times 12 = 123{,}200 \text{ in.-lb}$$
$$M_{u,\text{pos}} = 123{,}200 \times \tfrac{11}{16} = 84{,}700 \text{ in.-lb}$$

The moment capacity of the member is found by applying Eqs. (13-12) and (13-6).

$$\omega_{\max} = 0.6375 \times 0.85 \times \tfrac{87}{137} = 0.344$$
$$M_{u,\text{allow}} = 0.90 \times 12 \times 6^2 \times 3000 \times 0.344 \times 0.797 = 320{,}000 \text{ in.-lb}$$

The 7-in. slab is therefore adequate. The area of reinforcement is found by applying Eq. (13-7).

$$bdf_c = 12 \times 6 \times 2.55 = 183.6 \text{ kips}$$

$$\frac{2bf_c M_{u,neg}}{\phi} = \frac{2 \times 12 \times 2.55 \times 123.2}{0.90} = 8378 \text{ kips}^2$$

$$A_{s,neg} = \frac{183.6 - \sqrt{183.6^2 - 8378}}{50} = 0.488 \text{ sq in.}$$

Similarly, $\qquad\qquad A_{s,pos} = 0.328$ sq in.

The restriction $A_{s,min} = 200bd/f_y$ does not apply to a slab of uniform thickness, but the minimum steel area required for shrinkage and temperature stresses is 0.0020 times the gross concrete area in the present instance, with the maximum spacing of bars limited to 18 in. Then

$$A_{s,min} = 0.0020 \times 12 \times 7 = 0.168 \text{ sq in.} \qquad (a)$$

We shall use No. 4 bars. The area per bar is 0.20 sq in., and the required spacing for positive reinforcement is

$$12 \times \frac{0.20}{0.328} = 7.3 \text{ in.}$$

The spacing will be made 7 in., thus obtaining an equivalent steel area of 0.343 sq in. in a 12-in. width. Alternate bars will extend into the supports and the remaining bars will be bent. This arrangement satisfies Eq. (*a*) and the ACI requirement that "At least . . . one-fourth the positive moment reinforcement in continuous members shall extend along the same face of the member into the support. . . ." In the end regions, the trussed bars from the two adjoining spans will be supplemented with No. 4 straight bars 14 in. on centers, thus obtaining an equivalent steel area of 0.515 sq in.

We shall now locate the section at which the trussed bars can be bent upward. Figure 13-11*b* presents the bending-moment diagrams associated with the assumed values of maximum positive and maximum negative moment. Each diagram is a parabola having its summit at mid-span and having a total height of $\frac{1}{8}w_u L_n^2$. In accordance with Fig. 13-3 and Eq. (13-13), the capacity of the straight bars alone can be approximated conservatively by assuming that as A_s decreases the moment capacity of the member decreases in the same proportion. Therefore, the moment capacity of the straight bars alone is half the capacity of the total positive reinforcement. As a further simplification, the moment capacity of the total reinforcement will be taken as the actual maximum positive moment existing in the beam.

In Fig. 13-11b, let A denote the section at which the positive bending moment is half the maximum moment. Applying the principle of a parabola, that vertical offsets from the tangent through the summit vary as the squares of the corresponding horizontal distances, we obtain

$$\left(\frac{AM}{OM}\right)^2 = \frac{\frac{1}{32}}{\frac{1}{8}} = \frac{1}{4} \qquad \therefore AM = \tfrac{1}{2}(OM) = 3.50 \text{ ft}$$

The ACI Code states that "Reinforcement shall extend beyond the point at which it is no longer required to resist flexure for a distance equal to the effective depth of the member or 12 bar diameters, whichever is greater...." In the present instance, the required extension is 6 in., and this dimension will be applied to the straight part of the bar alone. Therefore, in Fig. 13-11a,

$$BM_{\min} = 3.50 + 0.50 = 4.00 \text{ ft}$$

We shall now locate the section at which the trussed bars can be bent downward. Two-thirds of the negative reinforcement continues beyond the bend point. Therefore, in Fig. 13-11b, let C denote the section at which the negative moment is two-thirds the maximum moment, or $\frac{2}{33}w_u L_n^2$. Then

$$\left(\frac{CM}{OM}\right)^2 = \frac{\frac{3}{88} + \frac{2}{33}}{\frac{1}{8}} = \frac{25}{33}$$

$$\therefore CM = 0.870(OM) = 6.09 \text{ ft}$$

In Fig. 13-11a,

$$DM_{\max} = 6.09 - 0.50 = 5.59 \text{ ft}$$

The horizontal projection of the bend is $7 - 1 - 1 = 5$ in., and therefore

$$BM_{\max} = 5.59 - 0.42 = 5.17 \text{ in.}$$

Adhering to the customary practice, we shall bend the bars upward at the fifth points, giving

$$OB = \tfrac{1}{5} \times 14 = 2.80 \text{ ft} \qquad BM = 7 - 2.80 = 4.20 \text{ ft}$$

The straight bars at the bottom must be checked for compliance with Eq. (13-21). In Fig. 13-11b, let E denote the point of inflection. Then

$$\left(\frac{EM}{OM}\right)^2 = \frac{\frac{1}{16}}{\frac{1}{8}} = \frac{1}{2} \qquad EM = 0.707(OM) = 4.95 \text{ ft}$$

At E,

$$V_u = 576 \times 4.95 = 2850 \text{ lb}$$

$$\frac{M_t}{V_u} + l_a = \frac{\frac{1}{2} \times 84{,}700}{2850} + 6 = 20.9 \text{ in.}$$

For the No. 4 bars,

$$l_d = 0.0004 \times 0.5 \times 50{,}000 = 10 \text{ in.} \qquad \text{OK}$$

The sections at which the top bars may be terminated are located by a similar procedure.

For the shrinkage and temperature reinforcement, use No. 3 bars $7\frac{1}{2}$ in. on centers, giving $A_s = 0.18$ sq. in.

14

REINFORCED-CONCRETE COLUMNS

14-1 Basic Design Concepts. If a concrete compression member has an unbraced length greater than three times the least lateral dimension, it must be supplied with longitudinal steel reinforcement. The member is said to be *spirally reinforced* if the reinforcement is secured in place by spiral hooping, and *tied* if the reinforcement is secured by means of intermittent lateral ties. We shall apply the strength method of analyzing reinforced-concrete members.

Since it is impossible in practice to obtain loading that is truly concentric, the ACI Code requires that a compression member be capable of sustaining a specified minimum bending moment at ultimate-load conditions. Consequently, every compression member must be viewed as a beam-column, notwithstanding the fact that the design load is assumed to be concentric.

A reinforced-concrete beam-column has three potential modes of failure: crushing of the concrete, which is assumed to occur when ϵ_c attains the value 0.003; yielding of the tension steel, which begins when its stress attains the value f_y; and the simultaneous crushing of the concrete and yielding of the tension steel.

We shall apply the following notational system:

A_g = gross area of section

h = overall depth of rectangular section normal to axis of rotation, or diameter of circular section

P_u = ultimate axial compressive load on member

P_o = allowable ultimate axial compressive load in absence of bending moment

P_b = axial compressive load in balanced ultimate-load system

M_u = ultimate bending moment in member

M_b = bending moment in balanced ultimate-load system

For every assigned value of P_u there is a corresponding value of M_u that causes impending failure, and vice versa. A set of values of P_u and M_u that causes impending failure constitutes the ultimate-load *system* acting on the member, and the basic problem in analyzing a given member may be regarded as that of identifying these systems. There is a particular set of values of P_u and M_u, denoted by P_b and M_b, respectively, that causes impending failure by the simultaneous crushing of the concrete and yielding of the tension steel. This set of values is referred to as the *balanced* ultimate-load system.

The ACI Code establishes the following design values:

a. For a spiral column, the minimum bending moment is that caused by an eccentricity of 1 in. or $0.05h$, whichever is greater, and $\phi = 0.75$.

b. For a tied column, the minimum bending moment is that caused by an eccentricity of 1 in. or $0.10h$, whichever is greater, and $\phi = 0.70$.

However, as the ratio of P_u to M_u diminishes, the member approaches a pure beam, and the ACI Code therefore allows the value of ϕ to approach 0.90 under certain specified conditions.

Assume that a member of rectangular section is subjected to bending with respect to a principal axis, and that the design moment exceeds the minimum moment set by the Code. The member must be investigated with respect to bending about the other principal axis, applying the specified minimum value. The two bending moments are considered to act separately.

14-2 Equations for Rectangular Members. Figure 14-1a is the cross section of a beam-column in which bending occurs about the x axis, inducing compression at edge a and tension at edge b. To allow for the loss of concrete area caused by the presence of the steel, we shall replace the true force in compression steel with an *effective* force by deducting the strength of the displaced concrete. Let

ϵ_A and ϵ_B = strain of reinforcement at A and B, respectively

f_A and f_B = stress in reinforcement at A and B, respectively

F_A and F_B = effective resultant force in reinforcement at A and B, respectively

$\quad\quad F_c$ = resultant force in concrete

d'_A and d'_B = distance from extreme edge to A and B, respectively

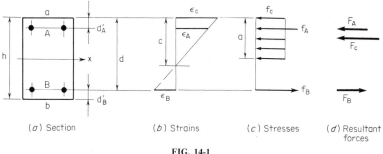

(*a*) Section   (*b*) Strains   (*c*) Stresses   (*d*) Resultant forces

FIG. 14-1

The sign convention is as follows: compression is positive, tension is negative. Thus, in the case illustrated in Fig. 14-1, F_A is positive and F_B is negative. Then

$$P_u = \phi(F_A + F_B + F_c) \tag{14-1}$$

If the steel stress is below the yield point, the strain diagram provides the following values of stress:

$$f_A = \epsilon_c E_s \frac{c - d'_A}{c} \qquad f_B = \epsilon_c E_s \frac{c - d}{c} \tag{14-2}$$

If ϵ_c has its limiting value of 0.003, these reduce to

$$f_A = 87,000 \frac{c - d'_A}{c} \qquad f_B = 87,000 \frac{c - d}{c} \qquad (14\text{-}2a)$$

By taking moments with respect to the x axis, we obtain the following result:

$$M_u = \phi \left[F_c \frac{h - a}{2} + F_A(h/2 - d'_A) - F_B(h/2 - d'_B) \right] \quad (14\text{-}3)$$

14-3   Interaction Diagrams.   As previously stated, the basic problem that arises in the analysis of a beam-column is to identify all sets of values of P_u and M_u that cause impending failure. A graph in which these sets of values are plotted is termed an *interaction diagram*. This graph therefore reveals the allowable value of M_u corresponding to a given value of P_u, and vice versa.

Example 14-1.   A short tied compression member having the cross section shown in Fig. 14-2 will be subjected to bending about the x axis, inducing compression at a and tension at b. The member is made of 3000-psi concrete and the steel is of Grade 40. Construct the interaction diagram for this member.

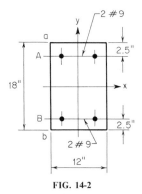

FIG. 14-2

Solution.   Refer to Fig. 14-3. Consider that we start with a truly concentric load equal to the ultimate-load capacity of the member. The entire concrete section is stressed to $0.85f'_c$ and the reinforcement is stressed to f_y in compression at both A and B. Now consider that we displace the load toward edge

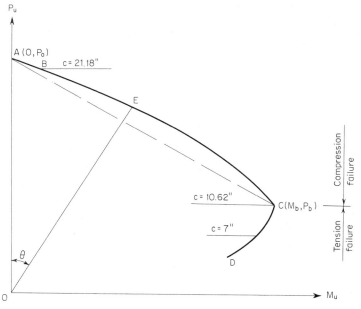

FIG. 14-3. Interaction diagram for beam-column.

a while reducing the magnitude of the load to its allowable value for the given eccentricity. The value of c decreases, and the behavior of the member passes through phases having the following boundary conditions:

CONDITION 1: $M_u = 0$

CONDITION 2: $a = h$ and therefore $c = h/\beta_1$

CONDITION 3: The member is subjected to the balanced ultimate-load system

It is necessary to investigate each of these boundary conditions in turn, and it is recommended that the reader draw the strain diagrams representing these conditions. The basic values are as follows:

$$f_c = 0.85f'_c = 0.85 \times 3000 = 2550 \text{ psi}$$
$$F_c = abf_c = a \times 12 \times 2550 = 30{,}600a$$

At *A* and *B*, $\qquad A_s = 2.00 \text{ sq in.}$

The minimum value of c at which $f_A = 40{,}000$ psi is found in this manner: By Eq. (14-2a),

$$f_A = 87{,}000 \, \frac{c - 2.5}{c} = 40{,}000 \qquad \therefore \ c = 4.63 \text{ in.}$$

To determine whether a reduction in ϕ is possible, we note that the reinforcement is symmetrically disposed and that

$$\frac{h - d'_A - d'_B}{h} = \frac{18 - 5}{18} > 0.70 \qquad \text{OK}$$

The value of P_u at which the reduction in ϕ begins is

$$P_u = 0.10f'_c A_g = 0.10 \times 3000 \times 18 \times 12 = 64,800 \text{ lb}$$

CONDITION 1

$$F_C = 30,600 \times 18 = 550,800 \text{ lb}$$
$$F_A = F_B = 2.00(40,000 - 2550) = 74,900 \text{ lb}$$
$$P_o = 0.70(2 \times 74,900 + 550,800) = 490,400 \text{ lb}$$

Theoretically, the minimum value of c associated with this condition is that at which the steel at B attains the value f_y. By Eq. (14-2a),

$$40,000 = 87,000 \frac{c - 15.5}{c} \qquad \therefore c = 28.69 \text{ in.}$$

CONDITION 2

$$a = 18 \text{ in.} \qquad c = 18/0.85 = 21.18 \text{ in.}$$

Since $c > 4.63$ in., $\qquad f_A = 40,000$ psi

$$f_B = 87,000 \frac{21.18 - 15.5}{21.18} = 23,300 \text{ psi}$$

$$F_c = 30,600 \times 18 = 550,800 \text{ lb}$$
$$F_A = 2.00(40,000 - 2550) = 74,900 \text{ lb}$$
$$F_B = 2.00(23,300 - 2550) = 41,500 \text{ lb}$$
$$P_u = 0.70(74,900 + 41,500 + 550,800) = 467,000 \text{ lb}$$
$$M_u = 0.70[6.5(74,900 - 41,500)] = 152,000 \text{ in.-lb}$$

CONDITION 3: By Eq. (14-2a),

$$f_B = 87,000 \frac{c - 15.5}{c} = -40,000 \text{ psi} \qquad c = 10.62 \text{ in.}$$

Since $c > 4.63$ in., $\qquad f_A = 40,000$ psi
$$a = 0.85 \times 10.62 = 9.03 \text{ in.} \qquad F_c = 30,600 \times 9.03 = 276,300 \text{ lb}$$
$$F_A = 74,900 \text{ lb} \qquad F_B = -80,000 \text{ lb}$$
$$P_b = 0.70(74,900 + 276,300 - 80,000) = 189,800 \text{ lb}$$
$$M_b = 0.70[276,300 \times 4.49 + 6.5(74,900 + 80,000)] = 1,573,000 \text{ in.-lb}$$

When 28.69 > c > 10.62 in., the member tends to fail in compression; when c < 10.62 in., it tends to fail by yielding of the tension reinforcement at B.

By assigning to c arbitrary values within the range from 10.62 to 28.69 in., we obtain other sets of values of P_u and M_u causing compression failure. As an illustration, set c = 14 in. Then

$$f_B = 87,000 \frac{14 - 15.5}{14} = -9320 \text{ psi}$$

$$a = 11.9 \text{ in.} \qquad F_c = 30,600 \times 11.9 = 364,100 \text{ lb}$$
$$P_u = 0.70(74,900 + 364,100 - 18,600) = 294,300 \text{ lb}$$
$$M_u = 0.70[364,100 \times 3.05 + 6.5(74,900 + 18,600)] = 1,203,000 \text{ in.-lb}$$

In a tension failure, the yielding of the tension steel causes a displacement of the neutral axis and a corresponding redistribution of stress until eventually a secondary failure occurs through crushing of the concrete. Therefore, when failure impends, the strain of the concrete is 0.003 and the strain of the tension steel exceeds that associated with initial yielding.

As an illustration of tension failure, set c = 7 in. Then

$$a = 5.95 \text{ in.} \qquad F_c = 30,600 \times 5.95 = 182,100 \text{ lb}$$
$$F_A = 74,900 \text{ lb} \qquad F_B = -80,000 \text{ lb}$$
$$P_u = 0.70(74,900 + 182,100 - 80,000) = 123,900 \text{ lb}$$
$$M_u = 0.70[182,100 \times 6.03 + 6.5(74,900 + 80,000)] = 1,473,000 \text{ in.-lb}$$

By plotting the points representing sets of values of P_u and M_u that cause impending failure and connecting them with smooth curves, we obtain the diagram shown in Fig. 14-3. The maximum value of M_u that the beam-column can sustain is that corresponding to the balanced ultimate-load system. As P_u assumes successively smaller values below P_b, the allowable ultimate moment decreases. The point at which the interaction diagram intersects the horizontal axis corresponds to the state at which the member functions as a pure beam, and the point at which the diagram returns to the vertical axis corresponds to the state at which the member functions as a pure tension member.

To simplify the calculations associated with a compression failure, the 1963 ACI Code allowed us to assume a linear relationship between P_u and M_u across the entire range of values in the compression-failure region. Geometrically, this simplification amounts to replacing the curve from A to C in Fig. 14-3 with a straight line connecting those points, and the equation of the straight line is

$$P_u = P_o - (P_o - P_b)\frac{M_u}{M_b} \tag{14-4}$$

Use of the straight line AC in lieu of the curve is conservative because it understates the allowable bending moment corresponding to a given value of P_u, and vice versa.

14-4   Eccentrically Loaded Columns.   Assume that the bending moment in a beam-column results from eccentricity of the load P_u, and let e denote this eccentricity. By setting $M_u = P_u e$ and rearranging terms, Eq. (14-4) is transformed to

$$P_u = \frac{P_o}{1 + (P_o - P_b)e/M_b} \tag{14-4a}$$

In Fig. 14-3, consider that we draw the radius vector to an arbitrary point E on the interaction diagram. Then

$$\tan \theta = \frac{M_u}{P_u} = \frac{P_u e}{P_u} = e$$

As we proceed along the interaction diagram from A to D, the value of c decreases and the value of e increases.

Example 14-2.   With reference to the member analyzed in Example 14-1, compute the eccentricity of loading associated with Conditions 2 and 3.
Solution.   Applying the results of the preceding calculations, we obtain the following:

CONDITION 2: $e = \dfrac{152,000}{467,000} = \textbf{0.33 in.}$

CONDITION 3: $e = \dfrac{1,573,000}{189,800} = \textbf{8.29 in.}$

Example 14-3.   With reference to the member analyzed in Example 14-1, compute the allowable ultimate load on this member as governed by bending about the x axis if the eccentricity with respect to that axis is (a) 4 in. and (b) 11 in. Equation (14-4a) may be applied where it is relevant.

Solution. As Example 14-2 discloses, the member fails in compression if $e < 8.29$ in. and in tension if $e > 8.29$ in.

PART a: From the calculations in Example 14-1,

$$P_o = 490,400 \text{ lb} \qquad P_b = 189,800 \text{ lb} \qquad M_b = 1,573,000 \text{ in.-lb}$$

By Eq. (14-4a),

$$P_u = \frac{490.4}{1 + 300.6 \times 4/1573} = \mathbf{277.9 \text{ kips}}$$

PART b: Assume $c > 4.63$ in. Then

$$F_A = 74.9 \text{ kips} \qquad F_B = -80 \text{ kips} \qquad F_c = 30.6a \text{ kips}$$

Applying Eqs. (14-1) and (14-3), we obtain

$$\frac{P_u}{\phi} = 74.9 + 30.6a - 80 = 30.6a - 5.1$$

$$\frac{M_u}{\phi} = 30.6a \left(\frac{18 - a}{2} \right) + 6.5(74.9 + 80)$$

$$= 275.4a - 15.3a^2 + 1006.9$$

Then

$$e = \frac{M_u}{P_u} = \frac{275.4a - 15.3a^2 + 1006.9}{30.6a - 5.1} = 11$$

giving

$$a^2 + 4a - 69.48 = 0$$

Solving,

$$a = 6.57 \text{ in.}$$

and the assumption concerning c is confirmed. Then

$$F_c = 30.6 \times 6.57 = 201.0 \text{ kips}$$

and

$$P_u = 0.70(74.9 + 201.0 - 80) = \mathbf{137.1 \text{ kips}}$$

This result can be verified by finding M_u.

$$M_u = 0.70[201.0 \times 5.72 + 6.5(74.9 + 80)] = 1510 \text{ in.-kips}$$

$$e = \frac{1510}{137.1} = 11.0 \text{ in.} \qquad \text{OK}$$

In design practice, it is necessary to compute the column capacity P_u by considering bending with respect to the y axis, applying the minimum eccentricities established by the Code.

Example 14-4. The member analyzed in Example 14-1 is to carry an ultimate load of 118 kips that is eccentric with respect to the x axis. Compute the maximum eccentricity with which the load may be applied.

Solution. By referring to the calculations in Example 14-1, it is seen that $\phi = 0.70$ in the present instance and that failure under the specified load would be one of tension. Assume again that $c > 4.63$ in. By Eq. (14-1),

$$P_u = 0.70(74.9 + F_c - 80) = 118$$
$$F_c = 173.7 \text{ kips} \quad \text{and} \quad a = 173.7/30.6 = 5.68 \text{ in.}$$
$$M_u = 0.70[173.7 \times 6.16 + 6.5(74.9 + 80)] = 1454 \text{ in.-kips}$$

$$e = \frac{1454}{118} = \textbf{12.32 in.}$$

Example 14-5. The member analyzed in Example 14-1 is to carry an ultimate load of 295 kips that is eccentric with respect to the x axis. Compute the maximum eccentricity.

Solution. Failure under the specified load would be one of compression. From Example 14-1,

$$P_o = 490.4 \text{ kips} \qquad P_b = 189.8 \text{ kips} \qquad M_b = 1573 \text{ in.-kips}$$

Equation (14-4a) yields

$$e = \frac{M_b(P_o - P_u)}{P_u(P_o - P_b)} = \frac{1573(490.4 - 295)}{295(490.4 - 189.8)} = \textbf{3.47 in.}$$

Applying the minimum eccentricities prescribed by the Code, it would be necessary in design practice to determine whether the member can support a load of this magnitude with an eccentricity about the y axis of $0.1 \times 12 = 1.2$ in.

The ACI has prepared an extensive set of interaction diagrams to aid the designer in selecting a column section to carry a load of given magnitude and eccentricity. These diagrams appear in the publication *Ultimate Strength Design Handbook*, Vol. 2, *Columns*.

15

COLUMN FOOTINGS

The columns and walls of a structure transmit their loads to the underlying soil through the medium of concrete footings, which serve to diffuse these loads over large areas and thus limit the soil pressure to an acceptable value. A reinforced-concrete footing supporting a single column under vertical loading differs from the usual type of flexural member in the following respects: The footing is subjected to bending in all directions, it carries a heavy load concentrated within a small area, and it is generally not reinforced in diagonal tension. As a result, the footing requires two-way reinforcement, its depth is usually governed by shear, the area of reinforcement may be subject to the lower limit imposed by the ACI Code, and the bearing pressure under the column may be critical if the footing is relatively narrow.

Since the upward reaction of the soil is collinear with the weight of the footing and of any overlying soil that is present, these weights do not contribute to the vertical shear and bending moment in the footing. Consequently, shearing and bending stresses are calculated by applying the *net* soil pressure, which is the pressure caused by the column load alone. It is convenient to interchange load and reaction mentally and visualize the footing as being subjected to an upward load transmitted by the underlying soil and a downward reaction supplied by the column. From this point of view, the footing functions as an overhanging beam.

There are two forms of vertical shear to be considered: that caused by the vertical column load, and that caused by the upward soil pressure on the overhanging portion of the footing. We shall refer to these as the *column-load shear* and the *soil-pressure shear*, respectively.

Since the ACI Code requires that a reinforced-concrete column be designed for a specified minimum eccentricity of loading, the footing supporting the column should also be designed for the resulting moment. However, this moment is often disregarded in office practice. We shall confine ourselves to the design of a con-centrically loaded column footing.

Figure 15-1 shows a square footing supporting a square, centrally located column under concentric vertical loading. The notational

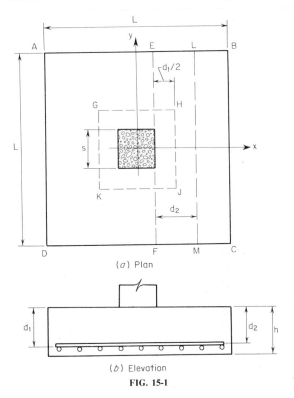

(a) Plan

(b) Elevation

FIG. 15-1

system is as follows:

P_u = ultimate column load

p_u = net soil pressure under ultimate load

A = area of footing in plan

L = length of side of footing

s = length of side of column

h = thickness of footing

f = bending stress

f_b = bearing stress under column

v_1 = nominal shearing stress caused by column-load shear

v_2 = nominal shearing stress caused by soil-pressure shear

V_{u1} = ultimate column-load shear

V_{u2} = ultimate soil-pressure shear

V_{u3} = ultimate shear on critical section for flexure

d_1 = effective depth of footing with respect to v_1

d_2 = effective depth of footing with respect to v_2 and f

b_1 = width of critical section with respect to v_1

Although the two layers of reinforcement have unequal effective depths, they are made alike for simplicity, and d_2 is taken as the effective depth of the upper layer, as indicated in Fig. 15-1. The depth d_1 may be taken to the horizontal plane between the layers. Overlapping effects caused by two-way reinforcement are ignored, and each layer of reinforcement is designed to resist alone the bending in its particular direction.

In accordance with the ACI Code, the critical section for v_1 is the surface formed by vertical planes through *GHJK*, which lie at a distance $d_1/2$ from the column faces. The critical section for v_2 is the vertical plane through *LM*, which is located at a distance d_2 from the column face. The critical section for f is the vertical plane through *EF* at the column face.

The depth of footing is usually governed by v_1, and it can be computed by means of the following equation:

$$(3.4v_1 + p_u)d_1{}^2 + s(3.4v_1 + 2p_u)d_1 = p_u(A - s^2) \qquad (15\text{-}1)$$

Example 15-1. A 20-in. square tied concrete column made of 4000-psi concrete and reinforced with four No. 8 bars carries a concentric service load of 525 kips, which consists of 215 kips dead load and 310 kips live load. Design a square footing of 3000-psi concrete and Grade 50 steel, using an allowable soil pressure of 4400 psf under service load.

Solution. The service load on the column is applied in establishing the area of the footing, but all other calculations are based on the ultimate-load conditions.

Assume that the footing weight is 9 percent of the column load. The required area of footing is

$$A = \frac{1.09 \times 525}{4.4} = 130.1 \text{ sq ft}$$

Set $L = 11$ ft 5 in. $= 11.42$ ft $A = 130.4$ sq ft
$$P_u = 1.4 \times 215 + 1.7 \times 310 = 828 \text{ kips}$$
$$p_u = 828/130.4 = 6.35 \text{ ksf}$$

The allowable shearing stresses as given by the ACI Code are

$$v_1 = 4\sqrt{f_c'} = 4\sqrt{3000} = 219 \text{ psi}$$
or
$$v_1 = 0.219 \times 144 = 31.54 \text{ ksf}$$
$$v_2 = 2\sqrt{f_c'} = 110 \text{ psi}$$

We shall apply Eq. (15-1) to obtain the thickness of footing, using kips and ft as the units. For example,

$$3.4v_1 + p_u = 3.4 \times 31.54 + 6.35 = 113.6 \text{ ksf}$$
Then
$$113.6d_1^2 + 199.9d_1 = 810.4$$
or
$$d_1^2 + 1.760d_1 = 7.134 \qquad \therefore d_1 = 1.932 \text{ ft}$$

Set $d_1 = 24$ in. $= 2.00$ ft. We shall test this value. In Fig. 15-1a,

$$GH = 1.67 + 2.00 = 3.67 \text{ ft}$$
$$V_{u1} = 6.35(130.4 - 3.67^2) = 742.3 \text{ kips}$$
or
$$V_{u1} = 828 - 6.35 \times 3.67^2 = 742.3 \text{ kips}$$
$$b_1 = 4 \times 3.67 = 14.68 \text{ ft}$$

$$v_1 = \frac{V_{u1}}{\phi b_1 d_1} = \frac{742.3}{0.85 \times 14.68 \times 2.00} = 29.7 \text{ ksf}$$

Reducing d_1 to 23 in. would make v_1 slightly excessive, and we therefore retain the value of 24 in. Allowing 3 in. for insulation and assuming the use of No. 8 bars, we obtain as the thickness $h = 24 + 1 + 3 = 28$ in. $= 2.33$ ft. Then

$$\text{Footing weight} = 130.4 \times 2.33 \times 0.150 = 45.6 \text{ kips}$$
$$\text{Assumed weight} = 0.09 \times 525 = 47.3 \text{ kips} \qquad \text{OK}$$

We shall now compute v_2.

$$d_2 = 24 - 0.5 = 23.5 \text{ in.}$$

In Fig. 15-1*a*,

$$LB = \tfrac{1}{2}(11.42 - 1.67) - 1.96 = 2.92 \text{ ft}$$
$$V_{u2} = 828 \times 2.92/11.42 = 211.7 \text{ kips}$$

or
$$V_{u2} = 6.35 \times 2.92 \times 11.42 = 211.7 \text{ kips}$$

$$v_2 = \frac{V_{u2}}{\phi L d_2} = \frac{211,700}{0.85 \times 137 \times 23.5} = 77 < 110 \text{ psi} \qquad \text{OK}$$

The footing size is therefore satisfactory, and it now remains to design the reinforcement. In Fig. 15-1*a*,

$$EB = 58.5 \text{ in.} = 4.88 \text{ ft}$$
$$V_{u3} = 828 \times 4.88/11.42 = 354 \text{ kips}$$
$$M_u = 354 \times \tfrac{1}{2} \times 58.5 = 10,360 \text{ in.-kips}$$

The steel area is found by applying Eq. (13-7).

$$bd_2 f_c = 137 \times 23.5 \times 2.55 = 8210 \text{ kips}$$

$$\frac{2bf_c M_u}{\phi} = \frac{2 \times 137 \times 2.55 \times 10,360}{0.90} = 8,043,000 \text{ kips}^2$$

$$A_s = \frac{8210 - \sqrt{8210^2 - 8,043,000}}{50} = 10.11 \text{ sq in.}$$

However, by the ACI Code,

$$A_{s,min} = \frac{200bd_2}{f_y} = \frac{200 \times 137 \times 23.5}{50,000} = 12.88 \text{ sq in.}$$

Use 17 No. 8 bars each way, giving $A_s = 13.4$ sq in. By Eq. (13-20), the development length is

$$l_d = 0.04 A_b f_y / \sqrt{f_c'} = 0.04 \times 0.79 \times 50,000/\sqrt{3000} = 28.8 \text{ in.}$$

Allowing 3 in. for insulation, it is seen that the bars extend beyond the section of maximum moment a distance of $58.5 - 3 = 55.5$ in., and their anchorage is therefore far more than is required by bond.

We shall now check the bearing stress at the interface of column and footing. Since the column is tied, the capacity of this member is based on a compressive stress in the concrete of

$$0.85 \phi f_c' = 0.85 \times 0.70 \times 4000 = 2380 \text{ psi}$$

The allowable bearing stress is

$$f_b = 2 \times 0.85 \phi f_c' = 2 \times 0.85 \times 0.70 \times 3000 = 3750 \text{ psi}$$

Therefore, the compression in the concrete of the column can be transmitted directly to the footing by bearing.

The compressive force in each steel bar of the column is transmitted to a dowel by proper splicing, and the dowel in turn transmits this force to the footing by bond. Assume there will be one No. 8 dowel of Grade 50 steel corresponding to each column bar. The development length for bars in compression is

$$l_d = 0.02 f_y d_b / \sqrt{f'_c} = 0.02 \times 50,000 \times 1.00 / \sqrt{3000} = 18.3 \text{ in.}$$

The footing can provide an embedment of $28 - 3 = 25$ in.

The footing is shown in Fig. 15-2.

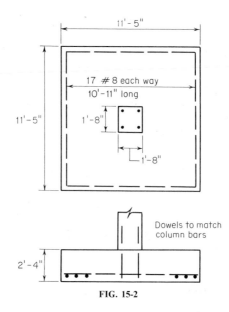

FIG. 15-2

16

FLUID MECHANICS

16-1 Buoyancy. A body immersed in a liquid is subjected to a vertical buoyant force equal to the weight of the displaced liquid. When this buoyant force becomes equal to the weight of the body, flotation results. Hence

$$V = \frac{W}{w} \qquad (16\text{-}1)$$

where V = volume of displaced liquid

W = weight of floating body

w = weight of unit volume of liquid

Example 16-1. A 10- by 12-in. timber 12 ft long is to be used as a buoy in salt water. What volume of concrete must be fastened to one end in order that 2 ft of the timber shall be above the water surface? Use the following weights: timber, 38 pcf; salt water, 64 pcf; concrete, 150 pcf.

Solution. Let x denote the volume of concrete.

$$V = 0.83 \times 1(12 - 2) + x = 8.3 + x$$
$$W = 0.83 \times 1 \times 12 \times 38 + 150x = 380 + 150x$$
$$8.3 + x = \frac{380 + 150x}{64}$$
$$x = \mathbf{1.76\ cu\ ft}$$

Example 16-2. A ship having a displacement of 24,000 tons and a draft of 34 ft in the ocean enters a harbor of fresh water. If the horizontal section of the ship at the water line is 32,000 sq ft, what depth of fresh water is required to float the ship? Assume that a marine ton is 2240 lb and that sea water and fresh water weigh 64 pcf and 62.2 pcf, respectively.

Solution. Let A denote the sectional area, d_1 the draft in the ocean, and d_2 the draft in fresh water.

$$d_2 - d_1 = \frac{V_2 - V_1}{A} = \frac{W}{A}\left(\frac{1}{62.2} - \frac{1}{64}\right) = \frac{0.00045W}{A}$$

$$= \frac{0.00045 \times 24,000 \times 2240}{32,000} = 0.76 \text{ ft}$$

$$d_2 = 34 + 0.76 = \textbf{34.76 ft}$$

Alternative Solution. If the draft remained 34 ft, the buoyant force in fresh water would be $W \times 62.2/64 = 0.972W$.

$$\therefore V_2 - V_1 = \frac{0.028W}{62.2} = 24,200 \text{ cu ft}$$

$$d_2 - d_1 = \frac{24,200}{32,000} = 0.76 \text{ ft} \qquad d_2 = 34.76 \text{ ft}$$

16-2   Bernoulli's Theorem.   In Fig. 16-1, liquid is flowing through a converging pipe from section 1 to section 2. For a frictionless flow,

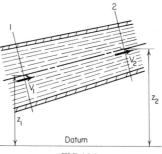

Datum

FIG 16-1

Bernoulli's theorem is expressed mathematically in this manner:

$$\frac{V_1^2}{2g} + \frac{p_1}{w} + z_1 = \frac{V_2^2}{2g} + \frac{p_2}{w} + z_2 \qquad (16\text{-}2)$$

where V = velocity of flow
$\quad\ g$ = acceleration due to gravity
$\quad\ p$ = liquid pressure
$\quad w$ = weight of unit volume of liquid
$\quad\ z$ = elevation above datum

The first term in each member of the equation represents the velocity head; the second, the pressure head; the third, the potential head.

Let Q denote the discharge (i.e., the volumetric rate of flow), A the sectional area, and D the pipe diameter. Since the flow is steady,

$$Q = A_1 V_1 = A_2 V_2 = 0.785 D_1^2 V_1 = 0.785 D_2^2 V_2$$

Let
$$h = \frac{p_1}{w} + z_1 - \frac{p_2}{w} - z_2$$

If V_1 is expressed in terms of V_2, Eq. (16-2) may be transformed to

$$V_2 = \sqrt{\frac{2gh}{1 - (D_2/D_1)^4}} \qquad (16\text{-}3)$$

To allow for the dissipation of energy caused by friction, let h_f denote the loss of head between the two sections. Equation (16-2) becomes

$$\frac{V_1^2}{2g} + \frac{p_1}{w} + z_1 = \frac{V_2^2}{2g} + \frac{p_2}{w} + z_2 + h_f \qquad (16\text{-}2a)$$

Similarly, let C be a friction coefficient representing the ratio of the true downstream velocity to that corresponding to frictionless flow under a head h. Equation (16-3) becomes

$$V_2 = C\sqrt{\frac{2gh}{1 - (D_2/D_1)^4}} \qquad (16\text{-}3a)$$

Example 16-3. A horizontal Venturi meter has a 24-in. upstream diameter and 16-in. throat diameter. The upstream pressure is 5.3 psi, and the pressure at the throat supports a water column 10 ft high. If the velocity coefficient is 0.98, how much water is flowing?

Solution. It is convenient to memorize this relationship: a pressure of 1 psi is equivalent to a water head of 2.31 ft.

$$\frac{p_1}{w} = 5.3 \times 2.31 = 12.24 \text{ ft} \qquad \frac{p_2}{w} = 10 \text{ ft}$$

$$h = 12.24 - 10 = 2.24 \text{ ft} \qquad \frac{D_2}{D_1} = \tfrac{16}{24} = 0.67$$

$$V_2 = 0.98 \sqrt{\frac{64.4 \times 2.24}{1 - 0.67^4}} = 13.14 \text{ fps}$$

$$Q = 0.785 \times 1.33^2 \times 13.14 = \textbf{18.34 cfs}$$

16-3   Power Associated with Fluid Flow.

The time rate of performing work, or the amount of work performed in a unit time, is called *power*. The energy possessed by a unit weight of a substance is called its *specific* energy. With respect to a liquid, the specific energy is the sum of the velocity, pressure, and potential heads.

Consider that a pump increases the specific energy of a flowing liquid by an amount E. Let Q denote the discharge and w the specific weight. The weight of liquid handled by the pump in a unit time is Qw, and the amount of energy that the pump delivers to the liquid in a unit time is QwE. Therefore, the power P delivered by the pump is

$$P = QwE \qquad (16\text{-}4)$$

A *horsepower* is 550 ft-lb/sec. Therefore, if the ft-lb-sec system of units is employed, the horsepower of the pump is

$$\text{Horsepower} = \frac{QwE}{550} \qquad (16\text{-}4a)$$

Example 16-4. A pump is discharging 8 cfs of water. Gages attached immediately upstream and downstream of the pump indicate a pressure differential of 36 psi. If the pump efficiency is 85 percent, what is the horsepower output and input of the pump?

Solution.   The work performed by the pump manifests itself as an increase in pressure head. Then

$$E = \frac{p_2}{w} - \frac{p_1}{w} = \frac{36 \times 144}{62.4} = 83.1 \text{ ft-lb/lb}$$

Let HP_o and HP_i denote the horsepower output and input, respectively (i.e., the power delivered *by* the pump and the power imparted *to* the pump). By Eq. (16-4a),

$$HP_o = \frac{8 \times 62.4 \times 83.1}{550} = \textbf{75.4}$$

$$HP_i = \frac{75.4}{0.85} = \textbf{88.7}$$

16-4   Flow through Orifices.   When liquid flows through an orifice and the upstream velocity is negligible, the discharge is

$$Q = CA\sqrt{2gh} \qquad\qquad (16\text{-}5)$$

where A = area of orifice

h = total head inducing flow

C = discharge coefficient, usually taken as 0.60 for a square-edged orifice in thin plate

Example 16-5.   Compute the discharge through a rectangular head gate opening (considered a square-edged orifice) 3 ft wide and 4 ft high. The water level on the upstream side is 15 ft and on the downstream side 6 ft above the top of the opening.

Solution

$$C = 0.60 \qquad h = 15 - 6 = 9 \text{ ft} \qquad A = 3 \times 4 = 12 \text{ sq ft}$$
$$Q = 0.60 \times 12\sqrt{64.4 \times 9} = \textbf{173.3 cfs}$$

16-5   Flow of Water over Weirs.   A *weir* is either a structure over which water flows or a notch in a structure in which water is confined. The *crest* of the weir is the bottom line of the issuing sheet of water. A weir having a sharp upstream edge is said to be *sharp-edged* (or *sharp-crested*). If the length of crest is less than the width of the structure, the weir is said to have *end contractions*; if these distances are identical, the weir is said to be *suppressed*.

The discharge over a sharp-edged suppressed rectangular weir in which the velocity of approach is negligible is given by the Francis formula as

$$Q = 3.33Lh^{3/2} \qquad \text{cfs} \qquad (16\text{-}6a)$$

where L = length of crest
h = head on weir, i.e., difference between elevation of crest and that of water surface upstream

An end contraction is equivalent in its effect to a reduction of the crest length by $0.1h$. Hence, for a sharp-edged rectangular weir having two end contractions, the discharge is

$$Q = 3.33(L - 0.2h)h^{3/2} \qquad \text{cfs} \qquad (16\text{-}6b)$$

Example 16-6. Compute the discharge over a sharp-edged rectangular weir 4 ft high and 10 ft long, with two end contractions, if the water in the canal is 4 ft 9 in. high. Disregard the velocity of approach.

Solution

$$Q = 3.33(10 - 0.2 \times 0.75)0.75^{3/2} = \textbf{21.3 cfs}$$

Example 16-7. Compute the proper length of a sharp-edged rectangular weir with one end contraction, over which 14 cfs is to flow, if the head on the weir will be 6 in. The velocity of approach may be neglected.

Solution

$$14 = 3.33(L - 0.1 \times 0.5)0.5^{3/2} \qquad \therefore \ L = \textbf{11.94 ft}$$

16-6 Flow of Liquid in Pipes.

For a circular pipe flowing full, it is convenient to express the loss of head due to friction in terms of the velocity head and the pipe diameter. The Darcy-Weisbach formula gives

$$h_f = f\,\frac{L}{D}\frac{V^2}{2g} \qquad (16\text{-}7)$$

where h_f = friction head in a pipe length L
f = friction coefficient
D = diameter of pipe

However, since h_f does not vary precisely as V^2/D in turbulent flow, the value of f is also a function of V and D. Values of f for several classes of pipe are given in Fig. 16-2.

In addition to the frictional loss, there is a loss of head at the pipe entrance, usually taken as half the velocity head, as well as other losses due to pipe bends, obstructions, etc.

In the ensuing problems, it is to be understood that the pipe is flowing full and that the flow is turbulent.

Example 16-8. Two reservoirs are connected by a 10,000-ft cast-iron pipe of 12-in. diameter. One reservoir is at elevation 550 ft, the other at elevation 450 ft. The coefficient for loss at pipe entry is 0.5, and the coefficient for friction loss is 0.018. Compute the discharge to the lower reservoir.

Solution. By considering a point near the pipe entrance and another at the terminus, we see that flow is induced by a head of 100 ft. This is consumed in the form of velocity head, friction head, and secondary losses.

$$h_f = 0.018 \left(\frac{10,000}{1}\right)\left(\frac{V^2}{2g}\right) = 180\frac{V^2}{2g}$$

$$100 = \frac{V^2}{2g} + 0.5\frac{V^2}{2g} + 180\frac{V^2}{2g} = 181.5\frac{V^2}{2g}$$

$$V = 5.96 \text{ fps} \qquad Q = 0.785 \times 1^2 \times 5.96 = \textbf{4.68 cfs}$$

Example 16-9. Two reservoirs are connected by a 7000-ft cast-iron pipe of 10-in. diameter. The difference in elevation of the water surfaces is 80 ft. Compute the discharge to the lower reservoir.

Solution. Since V and f are interdependent and both are unknown, the problem requires a solution by a trial-and-error procedure. Assume that V is 5 fps. Figure 16-2 indicates that f is 0.0204 for a 10-in. fairly smooth pipe.

$$h_f = 0.0204 \left(\frac{7000}{0.83}\right)\left(\frac{V^2}{2g}\right) = 171.4\frac{V^2}{2g}$$

$$80 = (1 + 0.5 + 171.4)\frac{V^2}{64.4} \qquad \therefore V = 5.46 \text{ fps}$$

Since the corresponding value of f is very close to the assumed value, the computed result is sufficiently precise.

$$Q = 0.785 \times 0.83^2 \times 5.46 = \textbf{2.97 cfs}$$

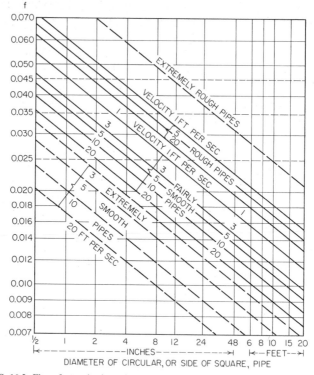

FIG. 16-2. Flow of water in pipes. (*From E. W. Schoder and F. M. Dawson,* Hydraulics, *McGraw-Hill Book Company, Inc., New York, 1934. By permission of the publishers.*)

Example 16-10. A pump drives water from a reservoir to a tank through 4000 ft of 8-in. cast-iron pipe at a rate of 2 cfs. The water surface in the tank is 95 ft above that in the reservoir. If the efficiency of the pump is 80 percent, what is the horsepower input of the engine?

Solution

$$Q = 2 \text{ cfs} \qquad V = 2/(0.785 \times 0.67^2) = 5.73 \text{ fps}$$

For a fairly smooth pipe, $f = 0.0212$.

$$\frac{p}{w} \text{ at pump} = \left[1 + 0.5 + 0.0212\left(\frac{4000}{0.67}\right)\right]\left(\frac{5.73^2}{64.4}\right) + 95 = 160.6 \text{ ft}$$

$$\text{Horsepower} = \frac{2 \times 62.4 \times 160.6}{0.80 \times 550} = \mathbf{45.6}$$

Example 16-11. A 10-in. pipe 9000 ft long leads from a pump to a reservoir having a water surface 200 ft above the pump. The flow is 3 cfs. Compute the pressure in the pipe at a point 3240 ft from the pump (measured along the pipe) and 110 ft above the pump, using $f = 0.024$. If the efficiency of the pump is 75 percent, what is the horsepower input?

Solution. Figure 16-3 is a graphical representation of Bernoulli's theorem, with the pump elevation as datum. Point A denotes the pump, B the pipe terminus, and C the given intermediate point. (The precise elevation of the pipe terminus is immaterial. For simplicity, it can be equated with that of the water surface in the reservoir.) CE represents the elevation and EF the pressure head at C; FG represents the friction head from A to C.

$$V = 3/(0.785 \times 0.83^2) = 5.55 \text{ fps}$$

$$HJ = h_f = 0.024 \left(\frac{9000}{0.83}\right)\left(\frac{5.55^2}{64.4}\right) = 124.5 \text{ ft}$$

By proportion,

$$FG = 124.5 \times 3240/9000 = 44.8 \text{ ft}$$
$$BJ = 200 + 124.5 = 324.5 \text{ ft}$$
$$EF = 324.5 - 110 - 44.8 = 169.7 \text{ ft}$$
$$p_c = 169.7/2.31 = \textbf{73.5 psi}$$
$$AD = 324.5 + 1.5 \times 5.55^2/64.4 = 325.2 \text{ ft}$$

$$\text{Horsepower} = \frac{3 \times 62.4 \times 325.2}{0.75 \times 550} = \textbf{147.6}$$

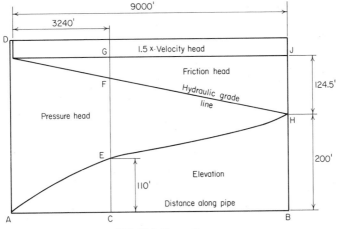

FIG. 16-3. Energy diagram.

Example 16-12. A 12-in. water line carrying 5 cfs branches into a 6-in. line 2000 ft long and an 8-in. line 1000 ft long. These branches rejoin and continue as a 12-in. line. Neglecting all secondary losses, compute the flow in each branch if f is 0.02 for all pipes.

Solution. Since the water pressure is the same in each branch at the point of juncture, it follows that the flow of 5 cfs is distributed in such manner that the friction heads in the branches are equal. Let the subscripts 1 and 2 refer to the 6- and 8-in. lines, respectively.

$$A_1 = 0.196 \text{ sq ft} \qquad A_2 = 0.349 \text{ sq ft}$$
$$0.196V_1 + 0.349V_2 = 5$$

$$h_f = 0.02 \left(\frac{2000}{0.5}\right)\left(\frac{V_1^2}{2g}\right) = 0.02 \left(\frac{1000}{0.67}\right)\left(\frac{V_2^2}{2g}\right)$$

$$\therefore V_1 = 0.612V_2$$

Substituting above,

$$V_2(0.196 \times 0.612 + 0.349) = 5$$
$$V_2 = 10.66 \text{ fps} \qquad V_1 = 0.612 \times 10.66 = 6.52 \text{ fps}$$
$$Q_1 = 0.196 \times 6.52 = \textbf{1.28 cfs} \qquad \text{in 6-in. pipe}$$
$$Q_2 = 0.349 \times 10.66 = \textbf{3.72 cfs} \qquad \text{in 8-in. pipe}$$
$$\Sigma Q = \overline{5.00} \text{ cfs} \qquad \text{OK}$$

Example 16-13. In Fig. 16-4, AM is a 10-in. pipe 8000 ft long, MB is an 8-in. pipe 5000 ft long, and MC is a 6-in. pipe 4000 ft long. Compute the discharge in each pipe, using $f = 0.02$.

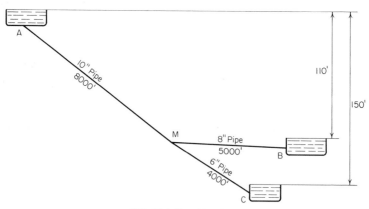

FIG. 16-4. Branching pipes.

Solution.   We shall disregard the velocity heads and all secondary losses. The friction head from A to B is 110 ft, and that from A to C is 150 ft. Let Q_1, Q_2, and Q_3 denote the discharge in AM, MB, and MC, respectively.

$$Q_1 = Q_2 + Q_3$$

By rearranging the terms in Eq. (16-7) and multiplying by the area, we obtain

$$Q = A \sqrt{\frac{2gD}{fL}} \sqrt{h_f}$$

If a denotes the friction head from A to M, then

Friction head from M to $B = 110 - a$
Friction head from M to $C = 150 - a$

The substitution of numerical values in the foregoing equation yields the following:

$$Q_1 = 0.545 \sqrt{\frac{64.4 \times 0.83}{0.02 \times 8000}} \sqrt{a} = 0.316 \sqrt{a}$$

Similarly

$$Q_2 = 0.229 \sqrt{110 - a} \qquad Q_3 = 0.124 \sqrt{150 - a}$$

To secure a first approximation, assume that $a = 70$ ft. This value gives

$$Q_1 = 2.64 \text{ cfs} \qquad Q_2 = 1.45 \text{ cfs} \qquad Q_3 = 1.11 \text{ cfs}$$
$$Q_2 + Q_3 = 1.45 + 1.11 = 2.56 < Q_1$$

The assumed value of a is therefore excessive. By continuing this process, we find that a is 68.0 ft, and the discharges are

$$Q_1 = \textbf{2.61 cfs} \qquad Q_2 = \textbf{1.49 cfs} \qquad Q_3 = \textbf{1.12 cfs}$$

16-7   Uniform Flow in Open Channels.   The *hydraulic radius* of a channel in which water is flowing is the ratio of the cross-sectional area to the wetted perimeter (WP) and is designated by R. There are several formulas that are widely used in computing uniform flow in a channel. The Chézy formula is

$$V = C\sqrt{Rs} \qquad \text{fps} \qquad (16\text{-}8)$$

where C = friction coefficient and usually lies between 50 and 150
$s = h_f/L$ = slope of hydraulic grade line = slope of channel

The Manning formula is

$$V = \frac{1.486}{n} R^{2/3} s^{1/2} \qquad \text{fps} \qquad (16\text{-}9)$$

where n = roughness coefficient.

The Chézy and Manning formulas are applicable to pipes as well as channels. Commonly used values of n are recorded in the accompanying table.

Surface	Value of n
Coated cast-iron pipe	0.013
Smooth lockbar and welded pipe	0.013
Riveted and spiral steel pipe	0.017
Vitrified sewer pipe	0.013
Glazed brickwork	0.013
Brick in cement mortar; brick sewers	0.015
Cement-mortar surfaces	0.013
Concrete pipe	0.015
Concrete-lined channels	0.016
Plank flumes, planed	0.012
Unplaned	0.013
With battens	0.015
Canals and ditches in straight and uniform earth	0.0225

The Kutter formula is also frequently used, but its complexity militates against its application in calculating R or s unless tables are available. The same values of n may be used in the Manning and Kutter formulas.

The hydraulically most efficient section for a circular channel is a semicircle; for a rectangular channel, a half square; for a trapezoidal channel, a half hexagon; etc.

Example 16-14. An 18-in. sewer pipe on a slope of 3 in 1000 is flowing half full. Compute the discharge, using $n = 0.013$.

Solution. For a semicircular water section,

$$R = \frac{A}{\text{WP}} = \frac{\pi D^2/8}{\pi D/2} = \frac{D}{4} = \frac{1.5}{4} = 0.375 \text{ ft}$$

$$V = (1.486/0.013) \times 0.375^{2/3} \times 0.003^{1/2}$$
$$= (1.486/0.013) \times 0.520 \times 0.055 = 3.26 \text{ fps}$$
$$Q = 0.5 \times 0.785 \times 1.5^2 \times 3.26 = \textbf{2.88 cfs}$$

Example 16-15. It is necessary to carry 1000 cfs of water from a dam to a power plant by means of a concrete-lined canal of rectangular cross section, 20 ft wide and 10 ft deep. If the canal is to flow full, what is the required fall of the water surface in feet per mile?

Solution. Use $n = 0.016$.

$$A = 20 \times 10 = 200 \text{ sq ft} \qquad \text{WP} = 20 + 2 \times 10 = 40 \text{ ft}$$
$$R = \tfrac{200}{40} = 5 \text{ ft} \qquad V = \tfrac{1000}{200} = 5 \text{ fps}$$

From Eq. (16-9),

$$s = \left(\frac{nV}{1.486R^{2/3}}\right)^2 = \left(\frac{0.016 \times 5}{1.486 \times 5^{2/3}}\right)^2 = 0.00034$$

Fall of water surface $= 0.00034 \times 5280 = \mathbf{1.80 \text{ ft per mile}}$

If we multiply both members of Eq. (16-9) by the area and rearrange terms, we obtain

$$AR^{2/3} = \frac{nQ}{1.486s^{1/2}} \qquad (16\text{-}10)$$

This equation is useful in computing the size of a channel when Q and s are known.

Example 16-16. In an irrigation project, it is necessary to deliver 68 cfs of water through a rectangular (half-square) concrete-lined flume from one canal to another that is 30 ft below and 200 ft distant. Determine the size of the flume.

Solution. Use $n = 0.016$. If b denotes the width of flume, the depth is $b/2$.

$$A = \frac{b^2}{2} \qquad \text{WP} = 2b$$

$$R = \frac{A}{\text{WP}} = \frac{b}{4} \qquad s = \tfrac{30}{200} = 0.15$$

$$AR^{2/3} = \frac{b^2}{2}\left(\frac{b}{4}\right)^{2/3} = \frac{nQ}{1.486s^{1/2}}$$

$$\therefore b^{8/3} = \frac{0.016 \times 68 \times 2 \times 4^{2/3}}{1.486 \times 0.15^{1/2}} \qquad b = 2.33 \text{ ft}$$

The flume should be 2 ft 4 in. wide and 1 ft 2 in. deep.

Example 16-17. Determine the depth (to the nearest foot) of a trapezoidal canal carrying water, using the following data: the flow is 800 cfs, the slope is 0.0004, the bottom width of the canal is 25 ft, and the slope of the sides is $1\frac{1}{2}$ horizontal to 1 vertical. Use $n = 0.014$.

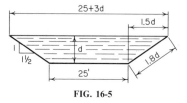

FIG. 16-5

Solution. See Fig. 16-5. If d denotes the depth,

$$A = \tfrac{1}{2}d(50 + 3d) \qquad \text{WP} = 25 + 3.6d$$

$$AR^{2/3} = \frac{0.014 \times 800}{1.486 \times 0.0004^{1/2}} = 377$$

Try $d = 5$ ft

$$A = 162.5 \text{ sq ft} \qquad \text{WP} = 43 \text{ ft} \qquad R = 3.78 \text{ ft}$$
$$AR^{2/3} = 162.5 \times 2.43 = 393$$

On trying $d = 4$ ft, we obtain $AR^{2/3} = 266$. The correct value of d is therefore 5 ft, or more precisely, **4.9 ft.**

16-8   Centrifugal Pumps.

In a centrifugal pump, liquid is conducted to the eye of an impeller by means of suction. The rotary motion of the impeller vanes causes the liquid to have a high velocity as it leaves the impeller periphery, and the induced velocity head is then converted to pressure head within the casing. The volume of liquid pumped per unit of time is termed the capacity or discharge, and the energy input per unit of time is termed the brake horsepower (bhp).

Let N denote the angular speed of the impeller, D the impeller outside diameter, Q the discharge, H the total dynamic head, and P the brake horsepower. Both N and D may be varied to meet the requirements of each situation. Consider first that D remains constant while the speed is changed from N_1 to N_2. If the efficiency is assumed to remain constant, it can be proved that

$$Q_2 = Q_1 \frac{N_2}{N_1} \qquad H_2 = H_1 \left(\frac{N_2}{N_1}\right)^2 \qquad P_2 = P_1 \left(\frac{N_2}{N_1}\right)^3 \quad (16\text{-}11)$$

On the other hand, if N remains constant while the diameter is changed from D_1 to D_2, then

$$Q_2 = Q_1 \frac{D_2}{D_1} \qquad H_2 = H_1 \left(\frac{D_2}{D_1}\right)^2 \qquad P_2 = P_1 \left(\frac{D_2}{D_1}\right)^3 \quad (16\text{-}12)$$

These relationships are referred to as the affinity laws.

The *specific speed* N_s of a pump is defined as the speed in rpm corresponding to a discharge of 1 gpm under a head of 1 ft. Its value is

$$N_s = N \frac{\sqrt{Q}}{H^{3/4}} \qquad (16\text{-}13)$$

The specific speed is a function of the ratio of the impeller eye diameter to the impeller outside diameter. In a multistage pump, the specific speed of a given stage is obtained by substituting the dynamic head pertaining to that stage.

Example 16-18.   PART *a*: A centrifugal pump delivers 700 gpm at 1800 rpm against a total dynamic head of 120 ft and requires 25 bhp for its operation. Compute the discharge, developed head, and brake horsepower if the speed is reduced to 1450 rpm. Assume that the efficiency remains constant.

PART *b*: Compute the capacity, head, and brake horsepower if the impeller of this pump is reduced in diameter from 9 to 8 in. and the speed is maintained at 1800 rpm.

PART *c*: If the pump, as first described, was designed as a two-stage pump, compute the specific speed.

Solution.   PART *a*

$$Q_1 = 700 \text{ gpm} \qquad H_1 = 120 \text{ ft} \qquad P_1 = 25 \text{ bhp}$$

$$\frac{N_2}{N_1} = \frac{1450}{1800} = 0.806$$

$$Q_2 = 700 \times 0.806 = 564 \text{ gpm}$$
$$H_2 = 120 \times 0.806^2 = 78.0 \text{ ft}$$
$$P_2 = 25 \times 0.806^3 = 13.1 \text{ bhp}$$

PART *b*

$$\frac{D_2}{D_1} = \frac{8}{9} = 0.889$$

$$Q_2 = 700 \times 0.889 = 622 \text{ gpm}$$
$$H_2 = 120 \times 0.889^2 = 94.8 \text{ ft}$$
$$P_2 = 25 \times 0.889^3 = 17.6 \text{ bhp}$$

PART *c*

$$N_s = 1800 \frac{\sqrt{700}}{60^{3/4}} = 2209 \text{ rpm}$$

17

SURVEYING AND ROUTE DESIGN

17-1 Horizontal Circular Curves. When the horizontal direction of a route changes, a curve is employed to effect the required transition. This curve is usually a circular arc that is tangent to the successive straight-line segments, as illustrated in Fig. 17-1. In railroad practice, the distance between two points on the curve is designated by the length of the chord joining these points, whereas in highway practice this distance is usually designated by the length of the intervening arc. The curvature of the circle is described by means of either the radius R or the central angle α subtended by a chord or an arc of 100-ft length. This angle is termed the *degree of curve*. If the chord length is 100 ft,

$$R = \frac{50}{\sin \frac{1}{2}\alpha} \qquad (17\text{-}1a)$$

If the arc length is 100 ft,

$$R = \frac{18{,}000}{\pi\alpha} \qquad (17\text{-}1b)$$

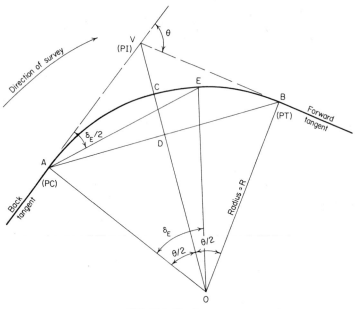

FIG. 17-1. Circular curve.

In Fig. 17-1, angle θ, which represents the change in direction of the route, is called the *intersection angle*; it is equal to the central angle AOB. Angle VAE is termed the *deflection angle* of station E with respect to tangent AV; it is equal to one-half the central angle AOE. The following notation is used to designate the various straight lines:

T = tangent distance = $AV = VB$
C = long chord = AB
M = middle ordinate = DC
E = external distance = CV

These quantities are related to R by the following equations:

$$T = R \tan \tfrac{1}{2}\theta \qquad (17\text{-}2)$$
$$C = 2R \sin \tfrac{1}{2}\theta \qquad (17\text{-}3)$$
$$M = R(1 - \cos \tfrac{1}{2}\theta) \qquad (17\text{-}4)$$
$$E = R \tan \tfrac{1}{2}\theta \tan \tfrac{1}{4}\theta \qquad (17\text{-}5)$$

In the field, the straight-line segments are extended to the PI (point of intersection), the intersection angle θ is measured, and after T has been computed, the PC (point of curve) of the curve is located. The transit is then set up at the PC, and each station is located by means of its deflection angle. For example, to locate station E in Fig. 17-1, the instrument is sighted along AE, the zero end of the tape is held at the preceding station, the chord distance from this station to E is set off on the tape, and the forward end of the tape is placed on the line of sight. (Where distances are expressed as arc lengths, the arc distance between successive stations must be translated to its equivalent chord distance for this purpose.) If all stations are visible from the PC, only one setup of the transit is required.

The PC and PT (point of tangency) of a curve generally do not occur at full stations. If the chord basis of measurement is used, we may for simplicity assume a proportionality between chord length and subtended angle.

Example 17-1. The PC of a circular curve of 1200-ft radius is at station $82 + 30$. The intersection angle is $28°$, and the chord distance between successive stations is to be 100 ft. Compute all data necessary to stake out the curve.

Solution. Let F represent the first station.

$$\sin \tfrac{1}{2}\alpha = \tfrac{50}{1200} = 0.04167$$
$$\therefore \tfrac{1}{2}\alpha = 2°23.3' = \text{incremental deflection angle}$$
$$AF = 70 \text{ ft} \qquad \therefore \delta_1 = 2°23.3' \times \tfrac{70}{100} = 1°40.3'$$
$$\text{Number of stations on curve} = 28°/(2 \times 2°23.3')$$
$$= 5.862$$
$$\text{Station at PT} = (82 + 30) + (5 + 86.2) = 88 + 16.2$$
$$\delta_7 - \delta_6 = 2°23.3' \times 16.2/100 = 0°23.2'$$

The deflection angles are recorded in the accompanying table.

Station	Deflection Angle
82 + 30	0
83	1°40.3'
84	4°03.6'
85	6°26.9'
86	8°50.2'
87	11°13.5'
88	13°36.8'
88 + 16.2	14°00'

Example 17-2. Explain the procedure for locating the PC and PT of a horizontal circular curve of known radius when the PI is inaccessible.

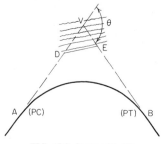

FIG. 17-2. Inaccessible PI.

Solution. See Fig. 17-2. Select accessible points D and E on the tangents, and measure the distance DE and angles DEV and EDV. θ equals the sum of these angles. Compute the tangent distance by Eq. (17-2). Applying the law of sines to triangle DEV, we obtain

$$DV = \frac{DE \sin DEV}{\sin \theta}$$

and

$$EV = \frac{DE \sin EDV}{\sin \theta}$$

$$AD = T - DV \qquad \text{and} \qquad BE = T - EV$$

The latter distances enable us to locate the PC and PT on their respective tangents.

Example 17-3. In Fig. 17-3, MN represents a straight railroad spur that intersects the curved highway route AB. Distances on the route are measured

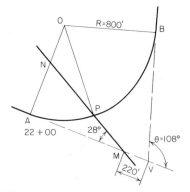

FIG. 17-3. Intersection of curve and straight line.

along the arc. Using the data shown, compute the station at the intersection point P.

Solution.   Draw line OP.

$$AV = T = 800 \tan 54° = 1101.1 \text{ ft}$$
$$AM = 1101.1 - 220 = 881.1 \text{ ft}$$
$$AN = AM \tan 28° = 468.5 \text{ ft}$$
$$ON = 800 - 468.5 = 331.5 \text{ ft} \qquad OP = 800 \text{ ft}$$
$$ONP = 90° + 28° = 118°$$

By the law of sines,

$$\sin OPN = \frac{\sin ONP \times ON}{OP} \qquad \therefore \ OPN = 21°27.7'$$

$$PON = 180° - (118° + 21°27.7') = 40°32.3'$$

$$\text{Arc } AP = \frac{2\pi \times 800 \times 40°32.3'}{360°} = 566.0 \text{ ft}$$

Station at $P = (22 + 00) + (5 + 66) = \mathbf{27 + 66}$

Example 17-4.   In Fig. 17-4, the horizontal curve AB has a radius of 720 ft and an intersection angle of 126°. The curve is to be realigned by rotating the forward tangent through an angle of 22° to the new position $V'B$ while B remains the PT. Compute the radius, and locate the PC of the new curve.

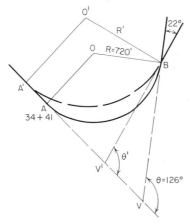

FIG. 17-4. Displacement of forward tangent.

Solution

$$\theta' = 126° - 22° = 104°$$
$$VB = 720 \tan 63° = 1413.1 \text{ ft}$$

Applying the law of sines to triangle $V'VB$, we obtain

$$V'B = \frac{1413.1 \sin 126°}{\sin 104°} = 1178.2 \text{ ft}$$

$$R' = \frac{1178.2}{\tan 52°} = \mathbf{920.5 \text{ ft}}$$

Similarly,

$$V'V = \frac{1413.1 \sin 22°}{\sin 104°} = 545.6 \text{ ft}$$

$$AV = VB = 1413.1 \text{ ft} \qquad A'V' = V'B = 1178.2 \text{ ft}$$
$$A'A = A'V' + V'V - AV = 310.7 \text{ ft}$$
$$\text{Station at new PC} = (34 + 41) - (3 + 10.7) = \mathbf{31 + 30.3}$$

Check.   Draw the long chords AB and $A'B$.

$$A'B = 2R' \sin \tfrac{1}{2}\theta' = 1450.7 \text{ ft}$$

In triangle $A'AB$,

$$AA'B = \tfrac{1}{2}\theta' = 52° \qquad A'AB = 180° - \tfrac{1}{2}\theta = 117°$$
$$ABA' = 180° - (52° + 117°) = 11°$$

By the law of sines,

$$A'A = \frac{1450.7 \sin 11°}{\sin 117°} = 310.7 \text{ ft} \qquad \text{OK}$$

On routes designed for relatively high speeds, a spiral is interposed between each straight-line segment and the main part of the circular curve to ease the transition from rectilinear to curvilinear motion, and vice versa.

17-2   Properties of Parabolic Routes.   The slope of a highway or railway route is referred to as its *grade*. When the terrain features necessitate a change in grade, this change is achieved gradually by connecting the successive straight-line segments with a smooth vertical curve. A parabolic arc is ideally suited for this purpose, for a reason that we will soon discuss. The length of the connecting curve is usually governed by the required sight distance corresponding to the design speed of the route, but other criteria may also be present.

Figure 17-5 is the projection of a highway route on a vertical plane. Consider that two straight-line segments having the indicated grades intersect at point *V* (the PVI). These segments are connected by a parabolic arc that starts at *A* (the PVC), terminates at *B* (the PVT), and has a summit at *S*. As shown in the drawing, the survey is assumed to proceed from left to right for orientation purposes.

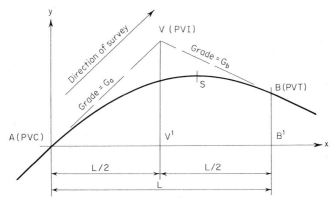

FIG. 17-5. Parabolic arc.

In the construction of a vertical parabolic curve, distances are measured horizontally and vertically. However, since the grades of the route are generally very small, no appreciable error accrues if distances measured along the curve are equated with their horizontal projections.

In Fig. 17-5, establish the horizontal and vertical coordinate axes shown, placing the origin at the PVC. Let

L = horizontal distance from PVC to PVT, referred to as the *length* of curve

G = grade at given station = dy/dx

r = rate of change of grade = $dG/dx = d^2y/dx^2$

The parabolic curve has the equation

$$y = ax^2 + bx \qquad (17\text{-}6)$$

where *a* and *b* are constants. By differentiating twice, we obtain

$$r = 2a \qquad (17\text{-}7)$$

Let V' and B' denote the projections of V and B, respectively, on the *x* axis. In Appendix A, it is demonstrated that

$$AV' = V'B' \qquad (17\text{-}8)$$

In Fig. 17-6, C and D are arbitrary points on the parabolic curve *AB*, and *CR* is the tangent to the curve at *C*. The vertical distance *DP* from *D* to the tangent through *C* is termed the *tangent offset* of *D* with respect to *C*. Let

G_c and G_d = grade of curve at C and D, respectively
$\quad G_{cd}$ = grade of chord *CD*
$\qquad t$ = tangent offset of *D* with respect to *C*

In Appendix A, it is demonstrated that

$$\Delta y = \left(\frac{G_c + G_d}{2}\right) \Delta x \qquad (17\text{-}9)$$

and
$$t = \tfrac{1}{2}r(\Delta x)^2 \qquad (17\text{-}10)$$

From Eq. (17-9), it follows that

$$G_{cd} = \frac{G_c + G_d}{2} \qquad (17\text{-}11)$$

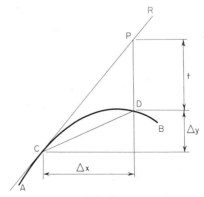

FIG. 17-6

In the special case where C lies at the origin, Eqs. (17-9) and (17-10) assume the following forms:

$$y = \left(\frac{G_c + G_d}{2}\right) x \qquad (17\text{-}9a)$$

$$t = \tfrac{1}{2}rx^2 \qquad (17\text{-}10a)$$

On the basis of the equations recorded above, it is possible to draw the following conclusions concerning the properties of a parabolic curve that conforms to Eq. (17-6):

1. From Eq. (17-7): The grade of the curve varies at a uniform rate along the horizontal. Therefore, the grade of the curve varies approximately at a uniform rate *along the curve*. It is this property that makes the parabola the logical choice of curve for effecting the change in grade of the route.

2. From Eq. (17-8): If tangents are drawn at two given points on the curve, these tangents intersect at a point that lies midway between the given points as measured along the horizontal.

3. From Eq. (17-10): Tangent offsets are directly proportional to the squares of their corresponding horizontal distances.

4. From Eq. (17-11): The grade of a chord is equal to the average of the grades of the curve at the two points joined by the chord.

Since the grade of the parabola varies at a uniform rate r, it follows that with reference to Fig. 17-6

$$G_d = G_c + r\,\Delta x \qquad (17\text{-}12)$$

Equation (17-6) has an important property that is very helpful in checking calculations. Consider that x assumes the successive values x_1, $x_2 = x_1 + h$, and $x_3 = x_1 + 2h$, where h is a constant. Let y_1, y_2, and y_3 denote the corresponding values of y. Then

$$(y_3 - y_2) - (y_2 - y_1) = 2ah^2 \qquad (17\text{-}13)$$

Therefore, if successive values of x differ by a constant amount, the *second differences* of the y values are all equal.

17-3   Design of Parabolic Routes.   Two methods of plotting a vertical parabolic curve are available: the average-grade method, which is based on Eq. (17-9a), and the tangent-offset method, which is based on Eq. (17-10a).

Example 17-5.   A grade of -5 percent is followed by a grade of $+1$ percent at station $38 + 50$ of elevation 374.50 ft. The change of grade is restricted to 2 percent in 100 ft. Compute the elevation of every 50-ft station on the curve, and locate the sag point (lowest point on the curve).

Solution

$$G_a - -5 \text{ percent} \qquad G_b = +1 \text{ percent}$$
$$r = 2 \text{ percent per 100 ft} = 1 \text{ percent per 50 ft}$$
$$= 0.02 \text{ percent per ft}$$

But
$$r = \frac{G_b - G_a}{L} \qquad \therefore L = \frac{1 - (-5)}{0.02} = 300 \text{ ft}$$

$$\text{Station at PVC} = \text{station at PVI} - \frac{L}{2} = 37 + 00$$

$$\text{Elevation of PVC} = \text{elevation of PVI} - G_a \frac{L}{2}$$

$$= 374.50 - (-0.05)150 = 382.00 \text{ ft}$$

Average-grade method.   The calculations are recorded in Table 17-1. The grade at the given station is found by adding 1 percent to the grade at the preceding station, and the grade shown in the fourth column is obtained by averaging the grade at the given station and that at the PVC. This average grade is then multiplied by x to obtain y, in accordance with Eq. (17-9a). The value of y is added to the elevation of the PVC to secure the elevation of the station.

TABLE 17-1   Calculations by Average-Grade Method

Station	x, ft	G	G_{avg}	y, ft	Elevation, ft
37 + 00	0	−0.05	−0.05	0	382.00
37 + 50	50	−0.04	−0.045	−2.25	379.75
38 + 00	100	−0.03	−0.04	−4.00	378.00
38 + 50	150	−0.02	−0.035	−5.25	376.75
39 + 00	200	−0.01	−0.03	−6.00	376.00
39 + 50	250	0	−0.025	−6.25	375.75
40 + 00	300	+0.01	−0.02	−6.00	376.00

As a means of testing the results, the second differences of the elevations have been generated in Table 17-2. The second differences are all equal, and the computed elevations are thus confirmed.

TABLE 17-2   Calculation of Second Differences

Elevations, ft	First differences, ft	Second differences, ft
382.00		
	2.25	
379.75		0.50
	1.75	
378.00		0.50
	1.25	
376.75		0.50
	0.75	
376.00		0.50
	0.25	
375.75		0.50
	−0.25	
376.00		

Let S denote the sag point and let the subscript s refer to this point. Since the grade at S is zero, we have, from Eq. (17-12),

$$G_s = G_a + rx_s = 0$$

$$\therefore x_s = -\frac{G_a}{r} = -\frac{-5}{0.02} = 250 \text{ ft}$$

Station at sag point $= (37 + 00) + (2 + 50) = 39 + 50$
Elevation of sag point $= 375.75$ ft (from Table 17-1)

In this example, the location of the sag point is evident from the fact that the elevations are symmetrical about station $39 + 50$, but this symmetry is concealed when the sag point does not fall at a full station.

Tangent-offset method. Let n denote the number of 50-ft stations from the PVC to the given station. Then $x = 50n$. Equation (17-10a) yields

$$t = 0.5 \times 0.0002x^2 = 0.0001x^2 = 0.0001(50n)^2$$

or
$$t = 0.25n^2 \tag{a}$$

The calculations are recorded in Table 17-3. The elevation of the tangent through the PVC increases by $(-0.05)50 = -2.50$ ft in 50 ft. The tangent offsets are then found by applying Eq. (a).

TABLE 17-3   Calculations by Tangent-Offset Method

Station	n	Elevation of tangent AV, + ft	Tangent offset, ft	=	Elevation, ft
37 + 00	0	382.00	0		382.00
37 + 50	1	379.50	0.25		379.75
38 + 00	2	377.00	1.00		378.00
38 + 50	3	374.50	2.25		376.75
39 + 00	4	372.00	4.00		376.00
39 + 50	5	369.50	6.25		375.75
40 + 00	6	367.00	9.00		376.00

Alternatively, tangent offsets can be found in this manner:

$$\text{Elevation of PVT} = \text{elevation of PVI} + G_b \frac{L}{2}$$

$$= 374.50 + 0.01 \times 150 = 376.00 \text{ ft}$$
$$\text{Tangent offset at PVT} = 376.00 - 367.00 = 9.00 \text{ ft}$$

Tangent offsets at intermediate stations are then obtained by proportion. For example, at station 39 + 00,

$$t = 9.00 \left(\frac{4}{6}\right)^2 = 4.00 \text{ ft}$$

In Example 17-5, the origin of coordinates was placed at the PVC in both methods of solution. However, if the PVC lies at a plus station rather than a full station, the calculations can be simplified by placing the origin at the first full station after the location of this station has been established.

In general, let M denote the station at the origin and G_m the grade at M. The parabolic curve can be plotted in a straightforward manner by applying the following equation, which stems directly from the tangent-offset method:

$$y = G_m x + \tfrac{1}{2}rx^2 \tag{17-14}$$

In the special case where the origin coincides with the summit or sag point, the equation of the curve reduces to

$$y = \tfrac{1}{2}rx^2 \tag{17-14a}$$

The design and plotting of a highway route can often be facilitated by use of a *grade diagram*. This is one in which distances along the horizontal base line represent horizontal distances along the route, and the ordinates (vertical distances from the base line) represent the grades at the corresponding stations. Since the grade of a parabola varies at a constant rate r, the grade diagram is a straight line having the slope r. The grade diagram is therefore extremely simple to draw and analyze. Let C and D denote two arbitrary stations on the route, and let y_c and y_d denote their respective ordinates. Then

$$y_d - y_c = \int_C^D \frac{dy}{dx}\, dx = \int_C^D G\, dx$$

$$= \text{area under grade diagram between } C \text{ and } D$$

Example 17-6. A grade of $+1.5$ percent intersects a grade of -2.5 percent at station $29 + 00$ of elevation 226.30 ft. The curve is to be 800 ft long. Locate the summit and calculate the elevation of the curve at station $32 + 00$, denoted by C.

Solution. It is understood that the left part of the route has the first grade specified and the right part has the second grade specified. Refer to Fig. 17-7a.

$$G_a = +1.5 \text{ percent} \qquad G_b = -2.5 \text{ percent}$$

$$r = \frac{G_b - G_a}{L} = \frac{-2.5 - 1.5}{800} = -0.005 \text{ percent per ft}$$

$$\text{Station at } A = \text{station at } V - \frac{L}{2} = 25 + 00$$

$$\text{Elevation of } A = 226.30 - 0.015 \times 400 = 220.30 \text{ ft}$$

Refer to Fig. 17-7b. By proportion,

$$\frac{as}{sb} = \frac{1.5}{2.5} \qquad \therefore \frac{as}{ab} = \frac{1.5}{1.5 + 2.5} = \frac{1.5}{4.0} = 0.375$$

$$as = 0.375 \times 800 = 300 \text{ ft}$$
$$\text{Station at } S = (25 + 00) + (3 + 00) = 28 + 00$$
$$y_s - y_a = \text{area } asm = \tfrac{1}{2} \times 300 \times 0.015 = 2.25 \text{ ft}$$
$$\text{Elevation of } S = 220.30 + 2.25 = \textbf{222.55 ft}$$

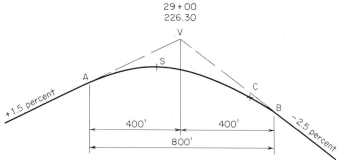

(a) Parabolic arc

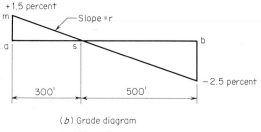

(b) Grade diagram

FIG. 17-7

Place the origin at S. At C,

$$x = 3200 - 2800 = 400 \text{ ft}$$

Applying Eq. (17-14a), we have

$$y = \frac{1}{2}\left(\frac{-0.005}{100}\right)400^2 = -4.00 \text{ ft}$$

Elevation of $C = 222.55 - 4.00 = \textbf{218.55 ft}$

Example 17-7. A $+5.2$ percent grade is followed by a -2.0 percent grade, the grades intersecting at station $22 + 30$ of elevation 194.60 ft. The parabolic curve is 900 ft long. Locate the station C on the curve that lies to the left of the summit and is at elevation 178.70 ft.

Solution. Refer to Fig. 17-8.

$$r = \frac{-2.0 - 5.2}{900} = -0.008 \text{ percent per ft}$$

Station at A = station at $V - \dfrac{L}{2} = 17 + 80$

Elevation of $A = 194.60 - 0.052 \times 450 = 171.20$ ft

$$as = \left(\frac{5.2}{5.2 + 2.0}\right) 900 = 650 \text{ ft}$$

Station at $S = (17 + 80) + (6 + 50) = 24 + 30$
$y_s - y_a = \text{area } asm = \frac{1}{2} \times 650 \times 0.052 = 16.90$ ft
Elevation of $S = 171.20 + 16.90 = 188.10$ ft

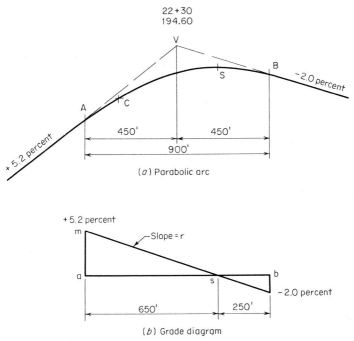

(*a*) Parabolic arc

(*b*) Grade diagram

FIG. 17-8

Place the origin at S. At station C,

$$y_c = 178.70 - 188.10 = -9.40 \text{ ft}$$

By Eq. (17-14a),

$$-9.40 = \frac{1}{2}\left(\frac{-0.008}{100}\right)x_c^{\,2}$$

Solving, $x_c = -484.8$ ft
Station at $C = (24 + 30) - (4 + 84.8) = \mathbf{19 + 45.2}$

Alternatively, x_c can be found by proportion, in this manner:

$$\left(\frac{x_c}{x_a}\right)^2 = \frac{y_c}{y_a} \qquad \therefore x_c = -650\sqrt{\frac{-9.40}{-16.90}} = -484.8 \text{ ft}$$

Example 17-8.   A vertical highway curve has an approach grade of -4.0 percent, and the PVC is at station $11 + 00$. The curve is to be 800 ft long and it is to have a turning point at station $16 + 00$ of elevation 86.0 ft. Compute the grade of the forward tangent, and locate the PVI.

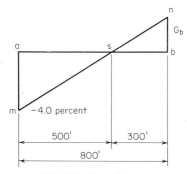

FIG. 17-9. Grade diagram.

Solution.   Let S denote the sag point. Refer to the grade diagram in Fig. 17-9.

At A: Station $= 11 + 00$
At S: Station $= 16 + 00$ Elevation $= 86.0$ ft
Then $as = 500$ ft   and $sb = 800 - 500 = 300$ ft

By proportion, $-\dfrac{G_a}{G_b} = \dfrac{500}{300}$ $\therefore G_b = \mathbf{+2.4 \text{ percent}}$

$$y_s - y_a = \text{area } asm = \tfrac{1}{2} \times 500(-0.04) = -10.0 \text{ ft}$$

Elevation of $A = 86.0 + 10.0 = 96.0$ ft

$$\text{Station at PVI} = \text{station at } A + \frac{L}{2} = \mathbf{15 + 00}$$

$$\text{Elevation of PVI} = \text{elevation of } A + G_a\frac{L}{2}$$

$$= 96.0 + (-0.04)400 = \mathbf{80.0 \text{ ft}}$$

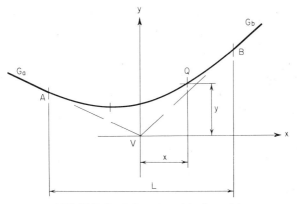

FIG. 17-10. Parabolic arc to contain given point.

A parabolic highway route must often be designed to contain some specific point in the vertical plane. For example, it may be necessary to effect a junction with an existing transverse road or to clear an existing object. Referring to Fig. 17-10, assume that the known data are G_a, G_b, the location of V, and the location of a point Q that is to lie on the parabolic curve. The problem is to determine the length of the particular curve that contains Q. Place the origin of coordinates at V, as shown, and let x and y denote the coordinates of Q. In Appendix A, it is demonstrated that

$$\frac{L + 2x}{L - 2x} = \sqrt{\frac{y - G_a x}{y - G_b x}} \tag{17-15}$$

This equation offers the simplest procedure for finding L.

Example 17-9. An ascending 4.8 percent grade intersects a descending 3.5 percent grade at station 82 + 60 of elevation 138.30. The highway is to pass over a culvert at station 80 + 53, and to provide ample clearance the elevation of the highway at this point must be at least 123.60 ft. If the length of curve is restricted to multiples of 100 ft, find the maximum length of curve that may be used. Verify the answer.

Solution. Assume initially that the elevation of the route will be precisely 123.60 ft at the specified station.

$$G_a = +4.8 \text{ percent} \qquad G_b = -3.5 \text{ percent}$$
$$x = 8053 - 8260 = -207 \text{ ft}$$
$$y = 123.60 - 138.30 = -14.70 \text{ ft}$$
$$G_a x = 0.048(-207) = -9.936 \text{ ft}$$
$$G_b x = (-0.035)(-207) = 7.245 \text{ ft}$$
$$y - G_a x = -14.70 - (-9.936) = -4.764 \text{ ft}$$
$$y - G_b x = -14.70 - 7.245 = -21.945 \text{ ft}$$

Substituting in Eq. (17-15),

$$\frac{L - 414}{L + 414} = \sqrt{\frac{-4.764}{-21.945}} = 0.4659$$

Solving, $$L = 1136 \text{ ft}$$

An increase in length depresses the curve, and a decrease in length elevates the curve. Therefore,

$$L_{\max} = \textbf{1100 ft}$$

Let Q denote station 80 + 53. To verify the solution, we shall compute the elevation of Q corresponding to a length of both 1136 ft and 1100 ft. Set $L = 1136$ ft, giving $L/2 = 568$ ft. Then

$$\text{Station at } A = (82 + 60) - (5 + 68) = 76 + 92$$
$$\text{Elevation of } A = 138.30 - 0.048 \times 568 = 111.036 \text{ ft}$$
$$\text{Horizontal distance from } A \text{ to } Q = 361 \text{ ft}$$

$$r = \frac{-3.5 - 4.8}{1136} = -0.00731 \text{ percent per ft}$$

Place the origin at A. By Eq. (17-14),

$$y_q = 0.048 \times 361 + \frac{1}{2}\left(\frac{-0.00731}{100}\right) 361^2 = 12.565 \text{ ft}$$
$$\text{Elevation of } Q = 111.036 + 12.565 = 123.60 \text{ ft}$$

Similarly, setting $L = 1100$ ft, we obtain

$$\text{Elevation of } Q = 123.91 \text{ ft}$$

17-4   Sight Distances.   Adequate visibility throughout its extent is a basic requirement of every highway route. The term *sight distance* is used to designate the length of road in front of motorists that lies within their vision. The minimum sight distances recommended by the American Association of State Highway Officials are presented in its publication "A Policy on Geometric Design of Rural Highways."

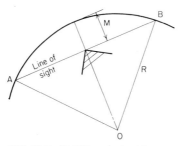

FIG. 17-11. Visibility on horizontal curve.

In Fig. 17-11, the horizontal circular arc AB represents the center line of the road. For simplicity, we may regard this as the path followed by the vehicle. Because of the obstruction that is present, an object at B remains invisible to motorists until they arrive at A. The radius of the curve must be sufficiently large to make the arc length AB at least equal to the required sight distance. Consequently, sight distance often serves as the criterion in establishing the curvature. Let S denote the sight distance, L the length of curve, R the radius, and M the middle ordinate of chord AB. From the geometry of the drawing, we obtain the following approximations:

If $S < L$,
$$R = \frac{S^2}{8M} \tag{17-16}$$

If $S > L$,
$$R = \frac{L(2S - L)}{8M} \tag{17-17}$$

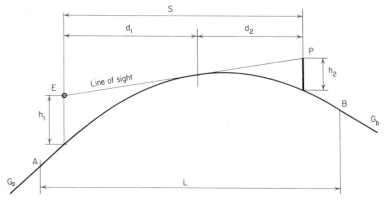

FIG. 17-12. Visibility on vertical summit curve.

Figure 17-12 is the profile of a vertical parabolic curve of length *L*. If an object occupies the indicated position to the right of the summit, its top point *P* becomes visible to motorists when their eyes are at *E*. Assume that the distance from *E* to *P* is the required sight distance *S*. Let h_1 denote the height of the observer's eye above the roadway and h_2 the height of object. (We may equate these distances with their vertical projections.) Consider the downward direction as positive, making G_b positive and G_a negative. Equation (17-10) for tangent offsets gives

$$d_1 = \sqrt{\frac{2h_1}{r}} \qquad d_2 = \sqrt{\frac{2h_2}{r}}$$

$$\therefore S = \frac{\sqrt{2h_1} + \sqrt{2h_2}}{\sqrt{r}}$$

Let *A* denote the change in grade from PC to PT, in percent. Then

$$r = \frac{G_b - G_a}{L} = \frac{A}{100L}$$

By substitution and simplification,

$$L = \frac{AS^2}{100(\sqrt{2h_1} + \sqrt{2h_2})^2} \qquad (17\text{-}18)$$

In like manner, when the sight distance exceeds the length of curve, we obtain

$$L = 2S - \frac{200(\sqrt{h_1} + \sqrt{h_2})^2}{A} \qquad (17\text{-}19)$$

Example 17-10. A vertical summit curve has tangent grades of $+2.6$ percent and -1.5 percent. Using $h_1 = 4.5$ ft and $h_2 = 4$ in., calculate the minimum length of curve for a sight distance of 450 ft.

Solution. Assume $S < L$. $A = 2.6 + 1.5 = 4.1$.

$$L = \frac{4.1 \times 450^2}{100(\sqrt{9} + \sqrt{0.67})^2} = \mathbf{570\ ft}$$

Since this value exceeds S, the use of Eq. (17-18) is justified.

There is no generally accepted standard for the design of a vertical sag curve. The AASHO recognizes four distinct criteria governing length of curve: headlight sight distance, rider comfort, drainage control, and a rule of thumb for general appearance.

17-5 Volumes of Earthwork. In order to compute the approximate volume of excavation or embankment entailed in constructing a road, the cross-sectional area of the cut or fill at each station along the route is determined. Two methods are available for calculating the volume of earthwork between successive stations: the average-end-area method and the prismoidal method.

Let A_1 and A_2 designate the areas of the sections at two consecutive stations, and L the intervening distance. When computed by the first method, the volume is

$$V = \frac{L}{2}(A_1 + A_2) \qquad (17\text{-}20)$$

The second method postulates that the earthwork between the stations is a prismoid. Consequently, each coordinate of the section lying midway between the end sections is the average of the corresponding coordinates at the ends. Let A_m designate the area of the center section. The formula for the volume of a prismoid is

$$V = \frac{L}{6}(A_1 + 4A_m + A_2) \qquad (17\text{-}21)$$

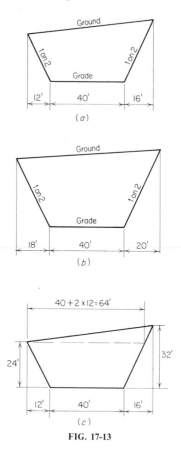

FIG. 17-13

Example 17-11. Figure 17-13*a* and *b* represents two highway cross sections 100 ft apart. Compute the volume of earthwork in cubic yards.

Solution. The coordinates of the first section are shown in Fig. 17-13*c*, and the coordinates of the other sections are similarly obtained.

$$A_1 = \frac{24(40 + 64) + (32 - 24)64}{2} = 1504 \text{ sq ft}$$

$$A_2 = \frac{36(40 + 76) + (40 - 36)76}{2} = 2240 \text{ sq ft}$$

$$A_m = \frac{30(40 + 70) + (36 - 30)70}{2} = 1860 \text{ sq ft}$$

$$V = \frac{100(1504 + 2240)}{2 \times 27} = 6933 \text{ cu yd}$$

or $$V = \frac{100(1504 + 4 \times 1860 + 2240)}{6 \times 27} = 6904 \text{ cu yd}$$

17-6   Plotting a Closed Traverse.   In Fig. 17-14, line *PQ* can be described by expressing its length *L* and its bearing α with respect to a reference meridian *NS*. The orthographic projection of *PQ* on the meridian is termed the *latitude*, and that on the parallel is termed the *departure*. Hence

$$\text{Latitude} = L \cos \alpha \qquad (17\text{-}22)$$
$$\text{Departure} = L \sin \alpha \qquad (17\text{-}23)$$

A positive latitude corresponds to a northerly bearing, and a positive departure to an easterly bearing.

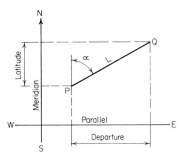

FIG. 17-14. Latitude and departure.

For a closed traverse, such as *abcde* in Fig. 17-15, the sum of the latitudes and the sum of the departures must equal zero. Manifestly, the lines can be arranged in some modified sequence, such as *adceb*, without destroying the closure.

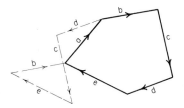

FIG. 17-15. Closure of traverse.

Example 17-12.   Complete the following table for a closed traverse:

Course	Bearing	Length
a	N32°27′E	110.8 ft
b		83.6
c	S8°51′W	126.9
d	S73°31′W	
e	N18°44′W	90.2

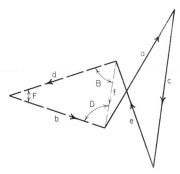

FIG. 17-16

Solution.   In Fig. 17-16, we introduce line f to form a closed traverse with the known courses a, c, and e. We then compute the following values:

Course	Latitude	Departure
a	+ 93.5 ft	+59.5 ft
c	− 125.4	− 19.5
e	+ 85.4	− 29.0
Total	+ 53.5	+11.0
f	− 53.5	−11.0

$$\tan \alpha_f = 11.0/53.5 \qquad \therefore \text{ Bearing of } f = \text{S}11°37′\text{W}$$

$$\text{Length of } f = \frac{53.5}{\cos \alpha_f} = 54.6 \text{ ft}$$

In triangle *fdb*,

$$B = \alpha_d - \alpha_f = 73°31' - 11°37' = 61°54'$$

To solve triangle *fdb*, apply the law of sines.

$$\sin F = \frac{f \sin B}{b} = \frac{54.6 \times \sin 61°54'}{83.6} \qquad \therefore F = 35°11'$$

$$D = 180° - (61°54' + 35°11') = 82°55'$$

$$d = \frac{b \sin D}{\sin B} = \frac{83.6 \times \sin 82°55'}{\sin 61°54'} = \mathbf{94.0 \ ft}$$

$$\alpha_b = 180° - (\alpha_d + F) \qquad \therefore \text{Bearing of } b = \mathbf{S71°18'E}$$

These results can be verified by selecting a new set of coordinate axes, one not parallel to the original set, and computing the corresponding latitudes and departures of all courses.

17-7 Calculation of Areas. The area of a tract encompassed within a closed traverse is frequently calculated by the double-meridian-distance method. In Fig. 17-17, the sum of m_1 and m_2 represents the *double meridian distance* (DMD) of line *AB*. Expressed algebraically,

$$\text{DMD}_{AB} = m_1 + m_2$$
$$\text{DMD}_{BC} = m_2 + m_3 = \text{DMD}_{AB} + m_3 - m_1 = \text{DMD}_{AB} + D_1 + D_2$$

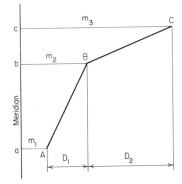

FIG. 17-17

In words, the DMD of a given course equals the sum of the following: the DMD of the preceding course, the departure of the preceding course, and the departure of the given course.

The area of trapezoid *ABba*, which we shall designate the *projection area* of *AB*, equals half the product of the DMD and latitude of the course. A projection area can be positive or negative. It can be proved that the area of a tract is numerically equal to the aggregate of the projection areas of its courses.

Example 17-13.   The balanced latitudes and departures of a closed transit and tape traverse are recorded in the accompanying table. Compute the area of the tract by the DMD method, using positive DMDs throughout.

Course	Latitude, ft	Departure, ft
AB	− 132.3	− 135.6
BC	+ 9.6	− 77.5
CD	+ 97.9	− 198.5
DE	+ 161.9	+ 203.6
EF	− 35.3	+ 186.7
FA	− 101.8	+ 21.3

Solution.   Consider that the meridian passes through *D*, since this is the most westerly point.

$$\text{DMD}_{DE} = 203.6 \text{ ft}$$
$$\text{DMD}_{EF} = 203.6 + 203.6 + 186.7 = 593.9 \text{ ft}$$
$$\text{DMD}_{FA} = 593.9 + 186.7 + 21.3 = 801.9 \text{ ft}$$

By continuing these calculations, we obtain the results shown in the table.

Course	Latitude	× DMD	= 2 × projection area
AB	− 132.3	687.6	− 90,970
BC	+ 9.6	474.5	+ 4,555
CD	+ 97.9	198.5	+ 19,433
DE	+ 161.9	203.6	+ 32,963
EF	− 35.3	593.9	− 20,964
FA	− 101.8	801.9	− 81,634
Total			− 136,617

$$\text{Area} = \tfrac{1}{2} \times 136,617 = \textbf{68,300 sq ft}$$

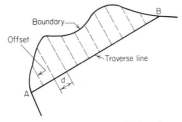

FIG. 17-18. Tract with irregular boundary.

Consider a tract of land having a curvilinear or meandering boundary, as shown in Fig. 17-18. To approximate the boundary, the perpendicular offsets from a straight line AB of the traverse are measured. Assume that the interval d between successive offsets is constant, and let $h_1, h_2, \ldots, h_n$ denote the offsets. We wish to compute the area lying between the traverse line AB and the irregular boundary. The *trapezoidal rule*, which is based on the assumption that the boundary is rectilinear between successive offsets, yields the equation

$$\text{Area} = d\left(\frac{h_1 + h_n}{2} + h_2 + h_3 + \cdots + h_{n-1}\right) \qquad (17\text{-}24)$$

Simpson's rule, which is based on the assumption that the boundary is the arc of a parabola between successive odd offsets, yields the equation

$$\text{Area} = \frac{d}{3}(h_1 + h_n + 2\Sigma h_{\text{odd}} + 4\Sigma h_{\text{even}}) \qquad (17\text{-}25)$$

where n is an odd integer and Σh_{odd} excludes h_1 and h_n.

Example 17-14. Compute the area between a traverse line and a meandering boundary, using the following data, all in feet: The interval between offsets is 25 ft, and the offsets are 0, 2.9, 3.6, 7.1, 8.9, 6.4, 3.1, 1.4, and 0.

Solution. The trapezoidal rule yields

$$\text{Area} = 25(2.9 + 3.6 + 7.1 + 8.9 + 6.4 + 3.1 + 1.4) = \textbf{835 sq ft}$$

Simpson's rule yields

$$\text{Area} = \tfrac{25}{3}[2(3.6 + 8.9 + 3.1) + 4(2.9 + 7.1 + 6.4 + 1.4)]$$
$$= \textbf{853 sq ft}$$

Now consider that the offsets are taken at irregular intervals, and let d_r denote the distance along the traverse line between the first and the rth offsets. If the boundary is assumed to be rectilinear between successive offsets, the encompassed area can be computed by applying either one of the following equations:

$$\text{Area} = \tfrac{1}{2}[d_2(h_1 - h_3) + d_3(h_2 - h_4) + \cdots + d_{n-1}(h_{n-2} - h_n) + d_n(h_{n-1} + h_n)] \tag{17-26}$$

$$\text{Area} = \tfrac{1}{2}[h_1 d_2 + h_2 d_3 + h_3(d_4 - d_2) + h_4(d_5 - d_3) + \cdots + h_n(d_n - d_{n-1})] \tag{17-27}$$

Example 17-15.   The following offsets were taken from stations on a base line to a meandering stream. All data are in feet. Compute the encompassed area.

Station	0 + 00	0 + 25	0 + 60	0 + 75	1 + 10
Offset	29.8	64.6	93.2	58.1	28.5

Solution.   Equation (17-26) yields

$$\text{Area} = \tfrac{1}{2}[25(29.8 - 93.2) + 60(64.6 - 58.1) + 75(93.2 - 28.5) + 110(58.1 + 28.5)] = \textbf{6590 sq ft}$$

Equation (17-27) yields

$$\text{Area} = \tfrac{1}{2}[29.8 \times 25 + 64.6 \times 60 + 93.2(75 - 25) + 58.1(110 - 60) + 28.5(110 - 75)] = 6590 \text{ sq ft}$$

Since the second method involves only positive terms, it possesses a distinct advantage over the first method.

17-8   Differential Leveling.   The elevation of a given benchmark (BM) with reference to that of a preceding benchmark is established by determining the elevations of several convenient intermediate points, called *turning points* (TP). For example, to obtain the elevation of BM2 in Fig. 17-19, the instrument is set up at point $L1$, point C is selected as a turning point, and the rod readings AB and CD are recorded. The former represents the backsight (BS) of BM1, and the latter represents the foresight (FS) of TP1. The instrument is then set up at point $L2$, and readings CE and FG are recorded. The elevation of BD is considered the height of instrument (HI) with respect to the preceding point, i.e., BM1. If a and b designate two

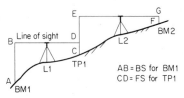

FIG. 17-19. Differential leveling.

successive turning points,

$$\text{Elevation}_a + \text{BS}_a = \text{HI}_a \qquad (17\text{-}28)$$
$$\text{HI}_a - \text{FS}_b = \text{elevation}_b \qquad (17\text{-}29)$$
$$\therefore \text{Elevation BM2} - \text{elevation BM1} = \Sigma\text{BS} - \Sigma\text{FS} \quad (17\text{-}30)$$

Example 17-16. Complete the following level notes. Show the usual arithmetic check. All data are in feet.

Point	BS	HI	FS	Elevation
BM42	2.076			180.482
TP1	3.408		8.723	
TP2	1.987		9.826	
TP3	2.538		10.466	
TP4	2.754		8.270	
BM43			11.070	

Solution. By applying the foregoing equations, we obtain the results shown in the accompanying table.

Point	BS	HI	FS	Elevation
BM42	2.076	182.558		180.482
TP1	3.408	177.243	8.723	173.835
TP2	1.987	169.404	9.826	167.417
TP3	2.538	161.476	10.466	158.938
TP4	2.754	155.960	8.270	153.206
BM43			11.070	144.890
Total	12.763		48.355	

Check. $144.890 - 180.482 = 12.763 - 48.355 = -35.592$

17-9   Stadia Surveying.   The location of selected control points for map plotting can be rapidly established by the process of stadia surveying, as illustrated in Fig. 17-20. The transit is set up over a reference point O, the rod is held at a control point N, and the telescope is sighted at a point Q on the rod. P and R represent the apparent locations of the stadia hairs on the rod. The rod intercept s, the vertical angle α, and the distance NQ are recorded in the notes, the last two usually being placed in the same column. When NQ is omitted, it is understood that this equals the height of instrument OM.

$$H = Ks \cos^2 \alpha + C \cos \alpha \qquad (17\text{-}31)$$
$$V = \tfrac{1}{2}Ks \sin 2\alpha + C \sin \alpha \qquad (17\text{-}32)$$
$$\text{Elevation of } N = \text{elevation of } O + OM + V - NQ \quad (17\text{-}33)$$

where $K = f/i =$ stadia interval factor $(= 100$ usually)
$\quad C = c + f =$ distance from center of instrument to principal focus

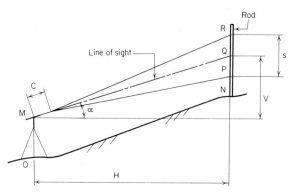

FIG. 17-20. Stadia surveying.

Example 17-17.   Complete the following stadia notes. The stadia constant is 100, and the value of $c + f$ is negligible. The instrument is at hub R (elevation 483.2 ft), and the height of the instrument is 5 ft.

Point	Rod intercept, ft	Vertical angle	Horizontal distance	Elevation
1	5.46	$+2°40'$ on 8 ft		
2	6.24	$+3°12'$ on 3 ft		
3	4.83	$-1°52'$ on 4 ft		

Solution. By applying the preceding equations, we obtain the following results:

Point	H, ft	V, ft	Elevation, ft
1	544.8	25.4	505.6
2	622.0	34.8	520.0
3	482.5	-15.7	468.5

17-10   Field Astronomy.   In Fig. 17-21, let P represent the position of Polaris at culmination at Greenwich (i.e., when Polaris appears to cross the Greenwich meridian) and M the simultaneous position of the mean sun. At the instant that Polaris appears to cross the observer's meridian at P', the mean sun is at M'. Since the apparent velocity of the mean sun is less than that of the stars, the distance h' is less than h, the difference being approximately 10 sec per hr of longitude. The distances h and h' represent, respectively, the time of culmination of Polaris at Greenwich and at the given site, measured from local noon. The computed local time of culmination must be

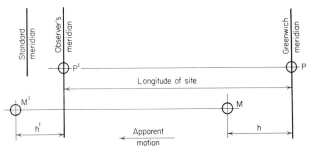

FIG. 17-21. Culmination of Polaris.

translated to watch time, which is the corresponding time at the standard meridian for that time zone.

Example 17-18. Determine Eastern Standard Time (75th meridian time) of upper culmination of Polaris at a place where the longitude is 81°W of Greenwich. Reference to an almanac reveals that Greenwich Civil Time (GCT) of upper culmination for the date of observation is $3^h20^m05^s$.

Solution. Since 15° of longitude corresponds to 1 hr,

$$\text{Longitude of site} = 81° = 5.4^h = 5^h24^m00^s$$
$$\text{Standard longitude} = 75° = 5^h$$

GCT of upper culmination at Greenwich.	$3^h20^m05^s$
Correction for longitude, 5.4 × 10 sec.	54^s
Local civil time of upper culmination at site	$3^h19^m11^s$
Correction to standard meridian.	24^m00^s
EST of upper culmination at site.	$3^h43^m11^s$ A.M.

The standard meridian is east of the observer's meridian, and the second correction must therefore be added.

In the astronomical triangle *PZS* in Fig. 17-22, *P* is the celestial pole, *Z* the observer's zenith, and *S* the apparent position of a star on the celestial sphere. Angle Z, which represents the azimuth of the star measured from the north, is computed by the equation

$$\tan^2 \tfrac{1}{2}Z = \frac{\sin (S - L) \sin (S - h)}{\cos S \cos (S - p)} \tag{17-34}$$

where *L* = latitude of site
h = altitude of star
p = polar distance = 90° − declination
$S = \tfrac{1}{2}(L + h + p)$

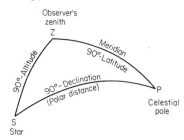

FIG. 17-22. Astronomical triangle.

An alternative method of solution is obtained by introducing an auxiliary angle M, which is defined as follows:

$$\cos^2 M = \frac{\cos p}{\sin h \sin L} \qquad (17\text{-}35)$$

Then $\qquad \cos(180° - Z) = \tan h \tan L \sin^2 M \qquad (17\text{-}36)$

Example 17-19. An observation of the sun was made at a latitude of 41°20′N. The altitude of the center of the sun, after correction for refraction and parallax, was 46°48′. By consulting a solar ephemeris, the declination of the sun at the instant of observation was found to be 7°58′N. What was the azimuth of the sun?

Solution. We shall first apply Eq. (17-34).

$$L = 41°20' \qquad h = 46°48' \qquad p = 90° - 7°58' = 82°02'$$
$$S = \tfrac{1}{2}(L + h + p) = 85°05'$$
$$S - L = 43°45' \qquad S - h = 38°17' \qquad S - p = 3°03'$$

$$
\begin{array}{ll}
\log \sin 43°45' = & 9.839800 \\
\log \sin 38°17' = & 9.792077 \\
\hline
 & 9.631877 \\
\end{array}
$$

$$
\begin{array}{ll}
\log \cos 85°05' = 8.933015 & \\
\log \cos 3°03' = 9.999384 & 8.932399 \\
\hline
2 \log \tan \tfrac{1}{2}Z \quad = & 0.699478 \\
\log \tan \tfrac{1}{2}Z \quad = & 0.349739 \\
\end{array}
$$

$$\tfrac{1}{2}Z = 65°55'03.5'' \qquad Z = \mathbf{131°50'07''} = \text{azimuth of sun}$$

Alternative Solution. The application of Eqs. (17-35) and (17-36) yields the following results:

$$
\begin{array}{lll}
\log \cos 82°02' & = & 9.141754 \\
\log \sin 46°48' & = 9.862709 & \\
\log \sin 41°20' & = 9.819832 & 9.682541 \\
\hline
2 \log \cos M & = & 9.459213 \\
\log \cos M & = & 9.729607 \\
\text{long} \sin M & = & 9.926276 \\
\hline
2 \log \sin M & = & 9.852552 \\
\log \tan 46°48' & = & 0.027305 \\
\log \tan 41°20' & = & 9.944262 \\
\hline
\log \cos(180° - Z) & = & 9.824119 \\
\end{array}
$$

$$Z = 131°50'07''$$

18

SOIL MECHANICS

18-1 Composition of Soils. In a three-phase soil mass, the voids, or pores, between the soil particles are occupied by moisture and entrapped air. The volumes of the ingredients of a soil mass are represented in Fig. 18-1. In a two-phase soil mass, the voids are occupied by moisture alone, and the soil mass is said to be fully saturated.

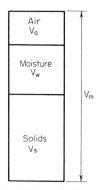

FIG. 18-1. Volumes of soil ingredients.

Let the subscripts s, w, a, and m refer to the solids, moisture, air, and soil mass, respectively. The following nomenclature and notational system are used:

$V = $ volume
$W = $ weight
$w = $ specific weight $= W/V$
$s = $ specific gravity
$n = $ porosity $= (V_w + V_a)/V_m$
$e = $ void ratio $= (V_w + V_a)/V_s$
$MC = $ moisture content $= W_w/W_s$
$S = $ degree of saturation $= V_w/(V_w + V_a)$

The weight of the mass is

$$W_m = V_s w_s + V_w w_w \qquad (18\text{-}1)$$

The weight of 1 cu cm of water is 1 gm. Therefore, if volume is expressed in cu cm and weight in gm, specific weight and specific gravity have the same numerical value, and Eq. (18-1) becomes

$$W_m = V_s s_s + V_w \qquad (18\text{-}1a)$$

The specific gravity of a soil mass is often referred to as its *apparent* specific gravity.

Example 18-1. A sample of moist soil having a volume of 98 cu cm and weighing 174 gm was allowed to dry out completely in an oven. If the final weight was 136 gm and the specific gravity of the solids is 2.68, what was the degree of saturation of the original mass?

Solution. For the initial conditions,

$W_w = 174 - 136 = 38$ gm $V_w = 38$ cu cm
$W_s = 136$ gm $V_s = 136/2.68 = 50.7$ cu cm
$\qquad\qquad V_a = 98 - (50.7 + 38) = 9.3$ cu cm
$\qquad\qquad S = 38/(38 + 9.3) = $ **80.3 percent**

Example 18-2. A soil sample weighing 122 gm has an apparent specific gravity of 1.82. The specific gravity of the particles is 2.53. After the sample is dried in an oven, the weight is 104 gm. Compute the volume of the solids and air in the original sample.

Solution. For the initial conditions,

$\qquad\qquad V_m = 122/1.82 = 67.0$ cu cm
$W_w = 122 - 104 = 18$ gm $V_w = 18$ cu cm
$W_s = 104$ gm $V_s = 104/2.53 = $ **41.1 cu cm**
$\qquad\qquad V_a = 67.0 - (41.1 + 18) = $ **7.9 cu cm**

Example 18-3. A soil mass has a porosity of 38 percent and degree of saturation of 92 percent. The specific gravity of the particles is 2.71. Find the specific weight of the soil mass in pcf.

Solution

$$n = 38 \text{ percent} \qquad S = 92 \text{ percent} \qquad s_s = 2.71$$

Set $V_m = 1$ cu ft. Then

$$V_w + V_a = 0.380 \text{ cu ft} \qquad V_s = 1 - 0.380 = 0.620 \text{ cu ft}$$

$$\frac{V_w}{V_w + V_a} = 0.92 \qquad V_w = 0.92 \times 0.380 = 0.350 \text{ cu ft}$$

$$W_m = (0.620 \times 2.71 + 0.350)62.4 = 127 \text{ lb}$$
$$w_m = \textbf{127 pcf}$$

It is often convenient to convert void ratio to porosity, or vice versa. The relationship is as follows:

$$n = \frac{e}{1 + e} \qquad \text{or} \qquad e = \frac{n}{1 - n} \tag{18-2}$$

Example 18-4. A soil mass has a void ratio of 0.82 and a moisture content of 26 percent. The particles have a specific gravity of 2.68. Find the degree of saturation.

Solution

$$e = 0.82 \qquad MC = 26 \text{ percent} \qquad s_s = 2.68$$
By Eq. (18-2),
$$n = 0.82/1.82 = 0.451$$

Set $V_m = 1$. Then

$$V_w + V_a = 0.451 \qquad V_s = 1 - 0.451 = 0.549$$

$$MC = \frac{W_w}{W_s} = \frac{V_w}{V_s s_s} = \frac{V_w}{0.549 \times 2.68} = 0.26$$

$$V_w = 0.383$$

$$S = \frac{V_w}{V_w + V_a} = \frac{0.383}{0.451} = \textbf{84.9 percent}$$

Example 18-5. A sample of fully saturated soil has a volume of 133 cu cm and weighs 242 gm. The particles have a specific gravity of 2.74. Compute the porosity, void ratio, moisture content, and specific weight of the soil in pcf.

Solution

$$V_m = V_s + V_w = 133 \text{ cu cm}$$
$$W_m = 2.74V_s + V_w = 242 \text{ gm}$$

Solving, $V_s = 62.6 \text{ cu cm}$ and $V_w = 70.4 \text{ cu cm}$

$$W_s = 2.74 \times 62.6 = 171.6 \text{ gm} \qquad W_w = 70.4 \text{ gm}$$
$$n = 70.4/133 = \textbf{52.9 percent} \qquad e = 70.4/62.6 = \textbf{1.125}$$
$$\text{MC} = 70.4/171.6 = \textbf{41.0 percent}$$
$$w_m = 242/133 = 1.820 \text{ gm/cu cm}$$

The specific weight of water is 1 gm/cu cm and 62.4 pcf. Therefore, to convert from the first system of units to the second we must multiply by 62.4. Then

$$w_m = 1.820 \times 62.4 = \textbf{114 pcf}$$

The upper surface of underground water is termed the *water table.* If a soil mass lies below the water table, it is fully saturated and acted upon by a buoyant force equal to the weight of the water displaced. If the soil mass occupies a unit volume, the buoyant force is equal to the specific weight of water. Therefore,

$$w_{m,\text{sub}} = w_{m,\text{sat}} - w_w \qquad (18\text{-}3a)$$

where $w_{m,\text{sat}}$ = specific weight of soil when saturated but not submerged

$w_{m,\text{sub}}$ = specific weight of soil when submerged

Expressing the volume of solids and moisture in terms of porosity, we obtain

$$w_{m,\text{sub}} = (s_s - 1)(1 - n)w_w \qquad (18\text{-}3b)$$

Applying Eq. (18-2), we then obtain

$$w_{m,\text{sub}} = \frac{(s_s - 1)w_w}{1 + e} \qquad (18\text{-}3c)$$

Example 18-6. A specimen of sand has a porosity of 35 percent, and the specific gravity of its solids is 2.70. Compute the specific weight of this soil in the submerged state, in pcf.

Solution. By Eq. (18-3b),

$$w_{m,\text{sub}} = 1.70 \times 0.65 \times 62.4 = \textbf{69.0 pcf}$$

Each type of soil is characterized by a maximum and minimum value of the void ratio. When the soil is in some intermediate state, its *relative density* (RD) is defined by the equation

$$RD = \frac{e_{\max} - e}{e_{\max} - e_{\min}} \qquad (18\text{-}4)$$

As the moisture in a clay mass exposed to the atmosphere evaporates, the soil passes through the following states, the boundaries of which are explicitly defined: liquid, plastic, semisolid, and solid. The moisture content corresponding to the transition from the liquid to plastic state is termed the *liquid limit*, that corresponding to the transition from plastic to semisolid state is termed the *plastic limit*, and that corresponding to the final transition is termed the *shrinkage limit*. The difference between the liquid and plastic limits is referred to as the *plasticity index*.

When the moisture content drops below the shrinkage limit, air enters the pores to occupy the space vacated by the escaping moisture, thereby keeping the volume of the mass constant throughout the solid state.

Example 18-7. A sample of saturated clayey soil in its natural state weighed 1920 gm and occupied a volume of 916 cu cm. After drying, the sample weighed 1530 gm and its total volume had diminished to 782 cu cm. The natural soil was found to have a liquid limit (LL) of 38.0 percent and a plastic limit (PL) of 19.4 percent. Compute the following: the original void ratio, the original moisture content, the specific gravity of the solids, the shrinkage limit (SL), the plasticity index (PI), and the weight of solids per cu ft of natural soil.

Solution. Figure 18-2 represents the volumes of the ingredients in the original state, at the shrinkage limit, and in the final state.

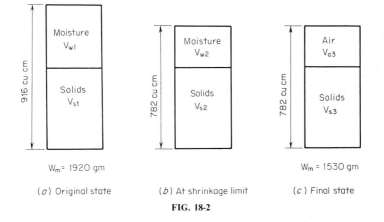

FIG. 18-2

The given and calculated values are recorded in Table 18-1, the given values being identified by an asterisk. We shall determine all values in this table for the sake of completeness, and we shall use the indicated designations as subscripts. It is recommended that the reader construct this table independently and record the calculated values as they emerge.

TABLE 18-1

State	Original	At shrinkage limit	Final
Designation	1	2	3
W_s, gm	1530	1530	1530
W_w	390	256	
W_m	1920*		1530*
V_s, cu cm	526	526	526
V_w	390	256	
V_a			256
V_m	916*	782*	782*

* Indicates given as opposed to calculated values.

The given values are as follows:

$$W_{m1} = 1920 \text{ gm} \qquad V_{m1} = 916 \text{ cu cm}$$
$$W_{m3} = 1530 \text{ gm} \qquad V_{m2} = V_{m3} = 782 \text{ cu cm}$$

Since the original mass is saturated,

$$V_{a1} = V_{a2} = 0$$

Since the final mass has no moisture,

$$V_{w3} = 0 \qquad W_{w3} = 0 \qquad W_{s3} = W_{m3} = 1530 \text{ gm}$$
$$\therefore W_{s1} = 1530 \text{ gm} \qquad \text{and} \qquad W_{s2} = 1530 \text{ gm}$$
$$W_{w1} = W_{m1} - W_{s1} = 1920 - 1530 = 390 \text{ gm}$$
$$V_{w1} = 390 \text{ cu cm} \qquad V_{s1} = 916 - 390 = 526 \text{ cu cm}$$
$$\therefore V_{s2} = 526 \text{ cu cm} \qquad \text{and} \qquad V_{s3} = 526 \text{ cu cm}$$
$$V_{w2} = 782 - 526 = 256 \text{ cu cm} \qquad W_{w2} = 256 \text{ gm}$$
$$V_{a3} = V_{w2} = 256 \text{ cu cm}$$
$$e_1 = 390/526 = \mathbf{0.741} \qquad MC_1 = 390/1530 = \mathbf{25.5 \text{ percent}}$$
$$s_s = 1530/526 = \mathbf{2.91}$$
$$SL = MC_2 = 256/1530 = \mathbf{16.7 \text{ percent}}$$
$$PI = LL - PL = 0.380 - 0.194 = \mathbf{18.6 \text{ percent}}$$

$$\frac{W_s}{V_{m1}} = \frac{1530}{916} \times 62.4 = \mathbf{104.2} \text{ lb of solids per cu ft of natural soil}$$

18-2   Transformation of Borrow Material.

The soil for an earth embankment is obtained by excavating soil from earth pits in the vicinity. These pits are referred to as *borrow pits*, and the excavated soil is termed *borrow material*. The soil that forms the embankment is called the *fill*.

There are two basic methods of obtaining a fill that meets the given specifications. If these specifications pertain to the grain size of the soil, it is necessary to mix the material obtained from two or more borrow pits, and the basic problem is to establish the relative proportions of the ingredients. If the specifications pertain to weight and moisture content, it may be possible to take material from one borrow pit alone and transform it to the fill by the addition of water and by compaction. In this case, the basic problem is to determine the relationship between the volume of borrow material and that of the fill, the amount of water to be added, and the degree of compaction.

Example 18-8. An earth embankment is to be constructed with soil having a specific weight of 115 pcf in the dry state and a moisture content of 16 percent. It has been found that the borrow material has a specific weight of 94 pcf and a moisture content of 9 percent, and its solids have a specific gravity of 2.70. Determine the following:

a. The volume of borrow material required to yield a unit volume of fill.

b. The number of gal of water to be added to each cu yd of borrow material, disregarding loss by evaporation.

Solution. All values are recorded in Table 18-2, the given and assumed values being identified by an asterisk. Although the volumes of air are superfluous in this example, they will nevertheless be included to make the table complete. It is recommended that the reader construct this table independently and record the calculated values as they emerge.

TABLE 18-2

	s	w, pcf	V, cu ft	W, lb	MC
Fill					
Solids	2.70		0.683	115.00	
Moisture			0.295	18.40	
Air			0.022		
Soil mass			1.000*	133.40	0.16*
Borrow					
Solids	2.70*		0.683	115.00	
Moisture			0.166	10.35	
Air			0.485		
Soil mass		94*	1.334	125.35	0.09*

* Indicates given and assumed values.

Let the subscripts 1 and 2 refer to the fill and borrow material, respectively. The given data are as follows:

$$MC_1 = 0.16 \qquad w_{m2} = 94 \text{ pcf} \qquad MC_2 = 0.09 \qquad s_{s2} = 2.70$$

$$\frac{W_{s1}}{V_{m1}} = 115 \text{ pcf}$$

Set $V_{m1} = 1$ cu ft, making $W_{s1} = 115.00$ lb.

$$\mathrm{MC}_1 = \frac{W_{w1}}{W_{s1}} = \frac{W_{w1}}{115.00} = 0.16 \qquad \therefore W_{w1} = 18.40 \text{ lb}$$

$$W_{m1} = 115.00 + 18.40 = 133.40 \text{ lb}$$
$$S_{s1} = S_{s2} = 2.70$$

$$V_{s1} = \frac{W_{s1}}{w_{s1}} = \frac{115.00}{2.70 \times 62.4} = 0.683 \text{ cu ft}$$

$$V_{w1} = \frac{18.40}{62.4} = 0.295 \text{ cu ft}$$

$$V_{a1} = 1 - (0.683 + 0.295) = 0.022 \text{ cu ft}$$
$$V_{s2} = V_{s1} = 0.683 \text{ cu ft} \qquad W_{s2} = W_{s1} = 115.00 \text{ lb}$$

$$\mathrm{MC}_2 = \frac{W_{w2}}{W_{s2}} = \frac{W_{w2}}{115.00} = 0.09 \qquad \therefore W_{w2} = 10.35 \text{ lb}$$

$$V_{w2} = \frac{10.35}{62.4} = 0.166 \text{ cu ft}$$

$$W_{m2} = 115.00 + 10.35 = 125.35 \text{ lb}$$

$$V_{m2} = \frac{W_{m2}}{w_{m2}} = \frac{125.35}{94} = 1.334 \text{ cu ft}$$

$$V_{a2} = 1.334 - (0.683 + 0.166) = 0.485 \text{ cu ft}$$

PART a

$$\frac{V_2}{V_1} = \frac{1.334}{1.000} = \textbf{1.334}$$

PART b: Let ΔV_w denote the volume of water to be added to the borrow material to obtain the fill. Then

$$\Delta V_w = 0.295 - 0.166 = 0.129 \text{ cu ft}$$
or
$$\Delta V_w = 0.129 \times 7.481 = 0.965 \text{ gal}$$
$$V_{m2} = 1.334/27 = 0.0494 \text{ cu yd}$$

$$\frac{0.965}{0.0494} = 19.5$$

Thus, **19.5** gal of water must be added to each cu yd of borrow material.

18-3 Design of Soil Mixtures.

Soil is often classified on the basis of its particle-size distribution, or *texture*. The texture is determined by a mechanical analysis that consists of passing a soil specimen through sieves of progressively smaller openings. The soil constituents are classified in the following manner: coarse aggregate, which is retained on a No. 10 sieve; fine aggregate, which passes a No. 10 sieve but is retained on a No. 200 sieve; and binder, which passes a No. 200 sieve.

In many instances, the specifications for a soil mass require that the fractional part by weight of the soil that passes a given sieve shall have a value lying within a stipulated *range* of values. For example, the specifications may state that the amount of material passing a No. 10 sieve shall lie between 40 and 60 percent by weight. It may be possible to meet the requirements by combining two types of materials. Let A and B denote these materials, and let

W_a, W_b, W_m = weight of A, B, and mixture, respectively

p_a, p_b, p_m = fractional part of weight of A, B, and mixture, respectively, that passes a given sieve

Then

$$W_a p_a + W_b p_b = W_m p_m = (W_a + W_b)p_m$$

$$\therefore p_m = \frac{W_a p_a + W_b p_b}{W_a + W_b} \tag{18-5}$$

Thus, p_m is a weighted average of p_a and p_b, and its value is therefore intermediate between the values of p_a and p_b. By setting $W_b = W_m - W_a$, we obtain the following equations for the relative weight of material A:

$$\frac{W_a}{W_m} = \frac{p_m - p_b}{p_a - p_b} = \frac{p_b - p_m}{p_b - p_a} \tag{18-6}$$

The subscripts a and b can be interchanged to give the relative weight of material B.

Example 18-9. A soil mass is to be formed by combining materials A and B. The texture of the ingredients and the required texture of the mix are recorded in Table 18-3.

   a. Calculate the maximum and minimum allowable value of the weight of material A corresponding to 1 lb of the mix.

   b. Selecting some allowable value of the weight of each ingredient, determine the weight of the mix passing the No. 40 sieve.

TABLE 18-3.   Sieve Analysis

| Sieve size | Weight passing, percent | | |
	Material A	Material B	Mix
$1\frac{1}{2}$ in.	100	100	100
$\frac{3}{4}$ in.	83	100	80 to 100
No. 4	72	90	50 to 85
No. 10	44	85	40 to 70
No. 40	9	64	20 to 50
No. 200	4	28	8 to 25

Solution. PART a: Set $W_m = 1$ lb. We shall calculate the values of W_a corresponding to the limiting values of p_m for each sieve size. However, we find by inspection that the lower limits of p_m for the $\frac{3}{4}$-in., No. 4, and No. 10 sieves cannot be satisfied because the given value of p_m does not lie between the corresponding values of p_a and p_b.

Table 18-4 presents the calculations of W_a in accordance with Eq. (18-6), the sieves being taken in reverse order. The lower and upper limits of p_m impose, respectively, upper and lower limits on W_a. The selected value of W_a must satisfy the limits imposed by *all* sieves. Taking the lowest value in the second column and the highest value in the third column of the table,

TABLE 18-4.   Weight of Material A

Sieve size	For lower limit of mix	For upper limit of mix
No. 200	$\dfrac{28 - 8}{28 - 4} = 0.833$	$\dfrac{28 - 25}{28 - 4} = 0.125$
No. 40	$\dfrac{64 - 20}{64 - 9} = 0.800$	$\dfrac{64 - 50}{64 - 9} = 0.255$
No. 10		$\dfrac{85 - 70}{85 - 44} = 0.366$
No. 4		$\dfrac{90 - 85}{90 - 72} = 0.278$
$\frac{3}{4}$ in.		$\dfrac{100 - 100}{100 - 83} = 0$

we obtain

$$W_{a,\text{max}} = \textbf{0.800 lb} \qquad W_{a,\text{min}} = \textbf{0.366 lb}$$

PART *b*: Set $W_a = 0.6$ lb and $W_b = 0.4$ lb. Equation (18-5) yields the following with respect to the No. 40 sieve:

$$p_m = 0.6 \times 0.09 + 0.4 \times 0.64 = \textbf{0.310 lb}$$

18-4 Shearing Capacity of Soil.   Failure of a soil mass is characterized by the sliding of one part past the other, and consequently a soil failure is one of shear. Resistance to sliding is provided by the cohesion of the soil and internal friction, but it is generally not possible to establish the precise manner in which resistance to shear failure is divided between the cohesion and the friction. If the shearing stress at a given point exceeds the cohesive strength, it is usually assumed for simplicity that the soil has mobilized its maximum potential cohesive resistance plus whatever frictional resistance is needed to prevent failure. The soil mass therefore remains in equilibrium if the ratio of the computed frictional stress to the normal stress is below the coefficient of internal friction of the soil.

Consider a soil prism in a state of triaxial stress. Let Q denote a point in this prism and P a plane that contains Q. Let

c = unit cohesive strength of soil
σ = normal stress at Q on plane P
σ_1 = maximum normal stress at Q
σ_3 = minimum normal stress at Q
τ = shearing stress at Q on plane P
θ = angle between P and plane on which σ_1 occurs
ϕ = angle of internal friction of soil

The stress analysis of the soil can be readily performed by constructing Mohr's circle of stress, which is developed in Art. 3-3. Refer to Fig. 18-3*a*. The shearing stress *ED* on plane *P* may be resolved into the cohesive stress *EG* and the frictional stress *GD*. We may therefore write

$$\tau = c + \sigma \tan \alpha$$

The maximum value of α that can exist at point Q is found by drawing the tangent *FH*.

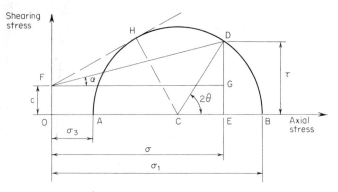

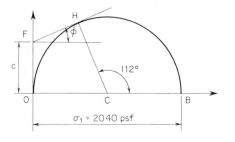

(a) Mohr's diagram for triaxial-stress condition

(b) Mohr's diagram for unconfined compression test

FIG. 18-3

Assume that failure impends at Q. Two conclusions may be drawn: The angle between FH and the base line OAB equals ϕ, and the angle between the plane of impending rupture and the plane on which σ_1 occurs equals one-half angle BCH.

The shearing capacity of a soil may be appraised by means of an *unconfined compression test*, in which a soil specimen is subjected to a vertical load without being restrained horizontally. The load is gradually increased until failure occurs. The stress σ_1 therefore occurs on a horizontal plane.

Example 18-10.   In an unconfined compression test on a soil sample, it was found that when the axial stress reached 2040 psf the soil ruptured along

In practice, three or four samples should be tested and the average value of ϕ and c determined.

18-5 Compressibility of Soil. When load is applied to a saturated soil mass, it is initially resisted solely by the moisture in the pores. However, since the induced moisture pressures are not uniform, a flow of moisture ensues, this flow resulting in a gradual reduction of the moisture content and the transfer of load from moisture to solids. It is this expulsion of moisture from the pores that causes the soil mass to contract, for the solids are assumed to be incompressible. As flow proceeds, the pressure differentials are reduced and the discharge continuously diminishes. In the following discussion, the soil is assumed to be saturated.

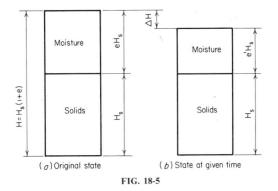

(a) Original state   (b) State at given time

FIG. 18-5

Let e and e' denote, respectively, the void ratio prior to vertical loading and at a given time after loading. Assume for convenience that the soil ingredients can be segregated into compartments, the thicknesses of which are represented in Fig. 18-5. The compression ΔH that has occurred up to the given time is $H_s(e - e')$, or

$$\Delta H = \frac{H(e - e')}{1 + e} \qquad (18\text{-}10)$$

Example 18-12. A building is to be constructed on a clay layer that overlies impervious rock, the thickness of the layer being 20 ft. In a laboratory test that simulated the field conditions, the void ratio of a specimen of this clay $\frac{3}{4}$ in. thick was reduced from 0.921 to 0.896. Compute the final settlement of the structure.

Solution

$$\Delta H = \frac{20 \times 12(0.921 - 0.896)}{1 + 0.921} = \textbf{3.1 in.}$$

Terzaghi and Peck offer the following equation for the compression of a confined stratum of normally loaded ordinary clay:

$$\Delta H = H \frac{C_c}{1 + e_0} \log_{10} \frac{p_f}{p_0} \qquad (18\text{-}11)$$

where H = thickness of stratum
C_c = compression index
e_0 = initial void ratio of clay
p_0 = initial soil pressure
p_f = final soil pressure
The compression index is a dimensionless property of the soil, and ΔH has the same units as H.

In calculating the compression of a soil mass, it is convenient to express the specific weight of a soil mass in terms of its void ratio and moisture content, the relationship being

$$w_m = \frac{s_s w_w \left[1 + (MC)\right]}{1 + e} \qquad (18\text{-}12)$$

If the soil mass is saturated, its void ratio is

$$e_{sat} = s_s(MC) \qquad (18\text{-}13)$$

If the mass is saturated but not submerged, its specific weight is

$$w_{m,sat} = \frac{s_s w_w[1 + (MC)]}{1 + s_s(MC)} \qquad (18\text{-}14)$$

Example 18-13.   The soil profile shown in Fig. 18-6 consists of a 36-ft layer of sand, a 20-ft layer of clay, and incompressible bedrock. The water table is 15 ft below the ground surface.

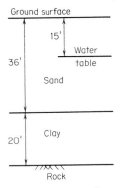

FIG. 18-6. Soil profile.

The sand has a void ratio of 0.52 and an average moisture content of 8 percent above the water table. The specific gravity of the sand particles is 2.63. The clay has an average moisture content of 30 percent and a compression index of 0.18. The specific gravity of the clay particles is 2.70.

A structure to be built on this soil will increase the soil pressure by 2100 psf. Calculate the compression of the layer of clay caused by this structure.

Solution. To determine the pressure of the clay, it is necessary to find the specific weight of the following: the sand above the water table, the sand below the water table, and the clay.

Sand above water table. By Eq. (18-12),

$$w_m = \frac{2.63 \times 62.4 \times 1.08}{1.52} = 116.6 \text{ pcf}$$

Sand below water table. By Eq. (18-3c),

$$w_{m,\text{sub}} = \frac{1.63 \times 62.4}{1.52} = 66.9 \text{ pcf}$$

Clay. Applying Eqs. (18-14), (18-3a), and (18-13), in that order, we obtain

$$w_{m,\text{sat}} = \frac{2.70 \times 62.4 \times 1.30}{1 + 2.70 \times 0.30} = 121.0 \text{ pcf}$$

$$w_{m,\text{sub}} = 121.0 - 62.4 = 58.6 \text{ pcf}$$
$$e_0 = 2.70 \times 0.30 = 0.81$$

The pressure on the clay varies across the depth of the layer. However, it will be sufficiently accurate to apply the mean pressure, which occurs at mid-depth. The initial soil pressure at this horizontal plane is caused by 15 ft of sand above the water table, 21 ft of sand below the water table, and 10 ft of clay. Then

$$p_0 = 15 \times 116.6 + 21 \times 66.9 + 10 \times 58.6 = 3740 \text{ psf}$$
$$p_f = 3740 + 2100 = 5840 \text{ psf}$$

$$\log_{10} \frac{p_f}{p_0} = 0.1935$$

By Eq. (18-11),

$$\Delta H = 20 \times 12 \times \frac{0.18}{1.81} \times 0.1935 = \mathbf{4.62 \text{ in.}}$$

19

WATER SUPPLY AND SEWERAGE

19-1 Hydraulics of Wells. Underground water constitutes an important source of water supply. The stratum of soil in which this water is present is known as an *aquifer*. On the basis of their hydraulic characteristics, wells are divided into two categories: *gravity* or *water-table* wells, and *artesian* or *pressure* wells. If the pressure at the surface of the surrounding underground water is atmospheric, the well is of the gravity type; if this pressure is above atmospheric because an impervious soil stratum overlies the aquifer, the well is artesian.

Assume that the water surrounding a well has a horizontal surface under static conditions. The lateral flow of water toward the well requires the existence of a hydraulic gradient, this gradient being caused by a difference in pressure. To create this difference in pressure, the surface of the surrounding water assumes the shape of an inverted "cone" during pumping of the well, as shown in profile in Fig. 19-1. This cone is known as the *cone of depression*, the cross section of the cone at the water surface is called the *circle of influence*, and the distance through which the water surface is lowered at the well is termed the *drawdown*. The discharge corresponding to a

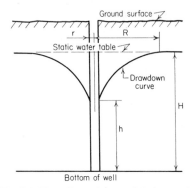

FIG. 19-1. Hydraulic gradient at well during pumping.

drawdown of 1 ft is called the *specific capacity* of the well. In accordance with Darcy's law, the velocity of flow to the pump is directly proportional to the hydraulic gradient, and the constant of proportionality is termed the *coefficient of permeability* of the aquifer.

Prior to construction of a well, there exists a state of equilibrium between the water entering the aquifer and the water discharged by the aquifer to rivers and other large bodies. In the *equilibrium* method of analysis, it is assumed that this equilibrium is not disturbed by action of the well. Consequently, the radius of the circle of influence corresponding to a given discharge through the well remains constant with respect to time. The yield of a well that completely penetrates the aquifer is given by the following set of equations:

Gravity well:
$$Q = \pi P \frac{H^2 - h^2}{\log_e (R/r)} \tag{19-1}$$

Artesian well:
$$Q = 2\pi P t \frac{H - h}{\log_e (R/r)} \tag{19-2}$$

where Q = volumetric rate of flow
P = coefficient of permeability of aquifer
t = thickness of aquifer
R = radius of circle of influence
r = radius of well

If H, h, and t are in ft and P is in gpm/sq ft, Q is in gpm.

As an approximation, R may be considered to be directly proportional to Q, and the foregoing equations assume the following forms:

Gravity well: $$Q = K \frac{H^2 - h^2}{\log_{10} CQ} \tag{19-1a}$$

Artesian well: $$Q = K' \frac{H - h}{\log_{10} C'Q} \tag{19-2a}$$

where C, C', K, and K' are constants for a given well.

Example 19-1. The depth of water at an artesian well was found to have the following values: 385 ft when the well was not pumped, 354 ft when the discharge was 400 gpm, and 318 ft when the discharge was 700 gpm. Applying Eq. (19-2a), determine what the discharge will be when the drawdown is 15 ft.

Solution. Common logarithms are used in the following calculations.

$$Q_1 = 400 \text{ gpm} \qquad H - h_1 = 385 - 354 = 31 \text{ ft}$$
$$Q_2 = 700 \text{ gpm} \qquad H - h_2 = 385 - 318 = 67 \text{ ft}$$

$$\log C'Q_1 = \log C' + \log 400 = \log C' + 2.6021$$
$$\log C'Q_2 = \log C' + 2.8451$$

Substituting in Eq. (19-2a),

$$400 = \frac{31K'}{\log C' + 2.6021} \tag{a}$$

$$700 = \frac{67K'}{\log C' + 2.8451} \tag{b}$$

Dividing Eq. (a) by Eq. (b) and rearranging,

$$\log C' = -1.5682 = 8.4318 - 10$$
$$\therefore C' = 0.0270$$

Substituting this value in Eq. (a) or Eq. (b) yields

$$K' = 13.3$$

When the drawdown is 15 ft, Eq. (19-2a) becomes

$$Q_3 = \frac{13.3 \times 15}{\log 0.0270Q_3}$$

or $$Q_3 \log 0.0270Q_3 = 199.5$$

By a trial-and-error solution,

$$Q_3 = \textbf{244 gpm}$$

19-2   Design of Storm Drains.   Storm drains for small areas are usually designed by the rational method, which consists of applying the following equation:

$$Q = Aci \qquad (19\text{-}3)$$

where Q = runoff, cfs
$\quad A$ = drainage area, acres
$\quad c$ = relative imperviousness of area
$\quad i$ = intensity of rainfall, in. per hr
The value of i diminishes as the storm continues; for simplicity, the value of c is assumed to remain constant.

Let t denote the duration of the storm in minutes. The relationship between i and t is given by Talbot as

$$i = \frac{105}{t + 15} \qquad (19\text{-}4)$$

Rainfall observations made in various cities have led to the development of many other formulas.

The *inlet time* of a drainage area is the maximum time required for a drop of water falling in this area to reach the sewer inlet. The *time of concentration* corresponding to a given point in the drainage system is the maximum time required for a drop of water falling in the contributing area to reach the given point.

Example 19-2.   Figure 19-2 records the area, relative imperviousness, and inlet time of every drainage district shown. Compute the runoff in each line of the storm sewer, assuming a velocity of 3 fps throught the system.

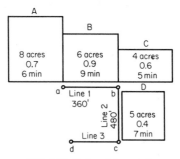

FIG. 19-2. Design of storm drain.

Solution. We shall compute the cumulative value of Ac for each line and apply Talbot's formula. The time required to cause maximum flow in a given line is assumed to be the time of concentration at the upstream inlet.

Line 1: A is the contributing area.

$$Ac = 8 \times 0.7 = 5.6 \qquad t_a = 6 \text{ min}$$
$$i = \tfrac{105}{21} = 5 \text{ in. per hr}$$
$$Q = 5.6 \times 5 = \textbf{28.0 cfs}$$

Line 2: A, B, and C are the contributing areas.

$$\Sigma Ac = 5.6 + 6 \times 0.9 + 4 \times 0.6 = 13.4$$
$$\text{Time of flow from } a \text{ to } b = 360/(3 \times 60) = 2 \text{ min}$$
$$\therefore \ t_b = 9 \text{ min} \qquad i = \tfrac{105}{24} = 4.38 \text{ in. per hr}$$
$$Q = 13.4 \times 4.38 = \textbf{58.7 cfs}$$

Line 3: A, B, C, and D are the contributing areas.

$$\Sigma Ac = 13.4 + 5 \times 0.4 = 15.4$$
$$\text{Time of flow from } b \text{ to } c = 480/(3 \times 60) = 2.7 \text{ min}$$
$$\therefore \ t_c = 11.7 \text{ min} \qquad i = 105/26.7 = 3.93 \text{ in. per hr}$$
$$Q = 15.4 \times 3.93 = \textbf{60.5 cfs}$$

19-3   Stabilization of Sewage.   When the organic matter in sewage has been decomposed to the simplest possible substances, the sewage is said to be *stabilized*. This effect can be achieved by using as agents either aerobic bacteria, which require oxygen, or anaerobic bacteria, which require an absence of oxygen. The first method is preferable. Aerobic stabilization comprises two successive stages, the first characterized by the oxidation of carbonaceous matter and the second by nitrification. For practical purposes, the sewage is considered to be stable when the first stage is 99 percent complete, a condition that is usually reached at the expiration of 20 days.

Aerobic decomposition is sustained by the dissolved oxygen in the liquid. A quantity of oxygen that is required or available is always expressed as the ratio of the amount of oxygen to the amount of liquid. If the ratio is by weight, the units are ppm (parts per million); if the ratio is the weight of oxygen to the volume of liquid, the units are mg/l (milligrams per liter). The numerical value is the same whichever system of units is used, for this reason: 1 liter equals 1000 cu cm. Therefore, 1 liter of fresh water weighs 1000 gm, or 1 million mg.

Example 19-3.   If water having a volume of 6000 gal contains 0.438 lb of oxygen, what is the oxygen content of the water in ppm?

Solution.   We shall apply the following relationship:

$$1 \text{ gal} = 0.1337 \text{ cu ft}$$
$$\text{Weight of oxygen} = 0.438 \text{ lb} = 438{,}000 \times 10^{-6} \text{ lb}$$
$$\text{Volume of water} = 6000 \times 0.1337 = 802.2 \text{ cu ft}$$
$$\text{Weight of water} = 802.2 \times 62.4 = 50{,}060 \text{ lb}$$

$$\frac{\text{Weight of oxygen}}{\text{Weight of water}} = \frac{438{,}000 \times 10^{-6}}{50{,}060} = \frac{8.75}{10^6} = \textbf{8.75 ppm}$$

Example 19-4.   If water having a volume of 15,000 gal has an oxygen content of 7.80 ppm, what is the weight of the oxygen in lb?

Solution.   We shall apply the following relationship directly: 1 gal of fresh water weighs 8.345 lb. Then

$$\text{Weight of oxygen} = 15{,}000 \times 8.345 \times 7.80 \times 10^{-6} = \textbf{0.976 lb}$$

The quantity of oxygen that is required to stabilize sewage is called its *biochemical oxygen demand* (BOD). Figure 19-3 depicts the manner in which BOD varies with time if the sewage is not exposed to the atmosphere. Let

t = elapsed time, days
L_a = initial BOD
L_t = BOD at expiration of t days
X_t = quantity of oxygen consumed in t days = $L_a - L_t$

As decomposition proceeds, oxygen is consumed at a rate directly

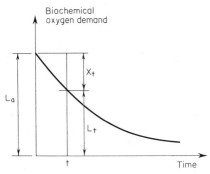

FIG. 19-3. **Variation of BOD with time.**

proportional to the amount of oxidizable matter that remains. Then

$$L_t = L_a \times 10^{-k_1 t} \qquad (19\text{-}5)$$
$$X_t = L_a(1 - 10^{-k_1 t}) \qquad (19\text{-}6)$$

In these equations, k_1 denotes the *coefficient of deoxygenation*. Its value varies with temperature. Let $k_{1(20)}$ and $k_{1(T)}$ denote the value of k_1 at 20°C and T°C, respectively. Then

$$k_{1(T)} = k_{1(20)} \times 1.047^{T-20} \qquad (19\text{-}7)$$

The expression "5-day BOD" actually means the quantity of oxygen consumed in 5 days, which is denoted X_5. This is the quantity that is usually determined in a laboratory test of a sewage specimen.

The maximum amount of oxygen that water can hold in suspension at a given temperature is called its *saturation content*. The difference between the saturation content and the true oxygen content is termed the *oxygen deficit*. Water tends to overcome this deficit by absorbing oxygen from the atmosphere. This process is known as *reaeration*, and the rate of reaeration is directly proportional to the magnitude of the deficit.

Consider that a stream exposed to the atmosphere becomes polluted at a given point. In Fig. 19-4, curve A depicts the variation

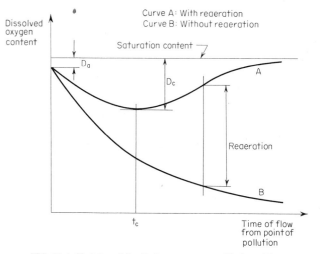

FIG. 19-4. Variation of dissolved oxygen content with time of flow.

of oxygen deficit with time of stream flow from the point of pollution. In addition to the notational system previously recorded, let

D_a = initial oxygen deficit
D_t = oxygen deficit at expiration of t days
D_c = critical (maximum) oxygen deficit
t_c = time corresponding to D_c, days
k_2 = coefficient of reaeration
f = self-purification constant = k_2/k_1

The equations of oxygen deficit are as follows:

$$D_t = \frac{k_1 L_a}{k_2 - k_1}(10^{-k_1 t} - 10^{-k_2 t}) + D_a \times 10^{-k_2 t} \qquad (19\text{-}8)$$

$$t_c = \frac{1}{k_1(f - 1)} \log_{10}\left\{ f\left[1 - (f - 1)\frac{D_a}{L_a} \right] \right\} \qquad (19\text{-}9)$$

$$D_c = \frac{L_a}{f} \times 10^{-k_1 t_c} \qquad (19\text{-}10)$$

Since it is imperative that the stream retain a certain minimum amount of dissolved oxygen, the critical oxygen deficit is of great significance. If the BOD of the sewage is such as to make this deficit excessive, the sewage must be treated to reduce its BOD before being discharged into the stream. The terms *influent* and *effluent* refer to the sewage entering and leaving the treatment plant, respectively. The mixture of discharged sewage and stream water is referred to as the *diluted sewage*.

Example 19-5.   At the point of discharge, diluted sewage has an oxygen deficit of 2.15 ppm, and the critical oxygen deficit downstream will be 4.19 ppm. If $k_1 = 0.1$ and $f = 1.8$, what is the allowable 5-day BOD of the diluted sewage at the point of discharge?

   Solution.   For the diluted sewage,

$$D_a = 2.15 \text{ ppm} \qquad D_c = 4.19 \text{ ppm}$$

A trial-and-error procedure is required to find L_a. Assume $L_a = 13.0$ ppm. Equation (19-9) yields $t_c = 2.421$ days, and Eq. (19-10) then yields $D_c = 4.136$ ppm, as compared with the given value of 4.19. A continuation of this procedure yields $L_a = 13.2$ ppm, to three significant figures. By Eq. (19-6),

$$X_5 = 13.2(1 - 10^{-0.1 \times 5}) = 13.2 \times 0.6838 = \mathbf{9.03 \text{ ppm}}$$

Again let Q denote the volumetric discharge and w the specific weight of water. Let the subscripts e, u, and d refer to the effluent from the treatment plant, the water upstream of the plant, and the diluted sewage, respectively. The total amount of oxygen consumed in the 5-day period immediately following discharge of the sewage is as follows:

$$Q_u w X_{5,u} + Q_e w X_{5,e} = Q_d w X_{5,d}$$

Eliminating w and setting $Q_d = Q_u + Q_e$ yields

$$Q_u X_{5,u} + Q_e X_{5,e} = (Q_u + Q_e) X_{5,d} \qquad (19\text{-}11)$$

Example 19-6.   A sewage-treatment plant for a residential city of 20,000 persons will receive an estimated sewage flow of 100 gal per capita per day, with a 5-day BOD of 0.17 lb per capita per day. The stream into which the sewage will be discharged has an average flow rate of 14 cfs and a 5-day BOD of 3 ppm. At the point of discharge, the diluted sewage will have an oxygen content of 8.50 ppm and a saturation content of 9.17 ppm. The diluted sewage must maintain a minimum oxygen content of 3.80 ppm, and tests indicate that this value is reached at the expiration of 2.60 days following discharge of the effluent. Using $k_1 = 0.1$ and $f = 1.75$, determine the relative amount by which the treatment plant must reduce the oxygen demand of the influent.

Solution.   The flow rate of the effluent and upstream water must be expressed in consistent units, and we shall use mgd (million gal per day) for this purpose. Before undertaking a numerical calculation, we shall indicate the manner in which units will be converted. The following relationships apply:

$$1 \text{ cu ft} = 7.481 \text{ gal}$$
$$1 \text{ gal of water weighs } 8.345 \text{ lb}$$

For influent

$$X_5 = \frac{\text{lb/person-day}}{(\text{gal/person-day})(\text{lb/gal})} = \frac{\text{lb}}{\text{lb}}$$

$$X_5 = \frac{0.17}{100 \times 8.345} = \frac{204}{10^6} = 204 \text{ ppm}$$

For influent and effluent

$$Q = \text{persons}(\text{gal/person-day}) = \text{gal/day}$$

$$Q = \frac{20{,}000 \times 100}{10^6} = 2.00 \text{ mgd}$$

For upstream water

$$Q = \frac{\text{cu ft}}{\text{sec}} \times \frac{\text{gal}}{\text{cu ft}} \times \frac{\text{sec}}{\text{hr}} \times \frac{\text{hr}}{\text{day}} = \frac{\text{gal}}{\text{day}}$$

$$Q = \frac{14 \times 7.481 \times 3600 \times 24}{10^6} = 9.05 \text{ mgd}$$

$$X_5 = 3 \text{ ppm}$$

For diluted sewage

$$D_a = 9.17 - 8.50 = 0.67 \text{ ppm}$$
$$D_c = 9.17 - 3.80 = 5.37 \text{ ppm}$$

By Eq. (19-10),

$$L_a = D_c f \times 10^{k_1 t_c} = 5.37 \times 1.75 \times 10^{0.1 \times 2.60} = 17.1 \text{ ppm}$$

By Eq. (19-6),

$$X_5 = 17.1(1 - 10^{-0.1 \times 5}) = 11.7 \text{ ppm}$$

For treatment system. Equation (19-11) yields

$$9.05 \times 3 + 2.00X_{5,e} = 11.05 \times 11.7$$

Solving,
$$X_{5,e} = 51 \text{ ppm}$$

$$\text{Relative reduction} = \frac{204 - 51}{204} = \textbf{75.0 percent}$$

Since the critical time was given, the value of D_a was superfluous.

APPENDIX A

ANALYSIS OF A PARABOLIC ARC

In Chap. 17, the basic properties of a parabolic arc are recorded and then applied in the design and plotting of a vertical highway curve. We shall now present the proof of these properties.

For our present purpose, it will be advantageous to utilize grade diagrams. The *grade* of a curve at a given point is the slope of the tangent to the curve at that point. As defined in Art. 17-3, the grade diagram of a curve is one in which distances along the horizontal base line represent horizontal distances in the original drawing, and vertical distances represent the grades at the corresponding points. Although grade diagrams are drawn in Art. 17-3 solely for parabolic arcs, these diagrams may be employed to analyze any type of curve whatever. We shall adhere to the notational system used in Arts. 17-2 and 17-3.

With reference to Fig. A-1*a*, *C* and *D* are arbitrary points lying on a parabolic arc having the equation

$$y = ax^2 + bx$$

where *a* and *b* are constants. The tangents to the curve at *C* and *D* intersect at *V*. The grade diagram appears in Fig. A-1*b*. In Arts.

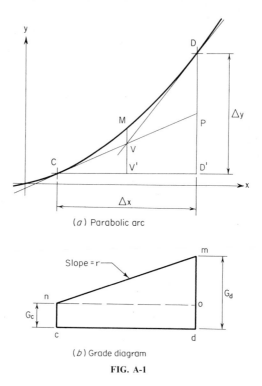

(*a*) Parabolic arc

(*b*) Grade diagram

FIG. A-1

17-2 and 17-3, it is proved that:

1. The grade of a parabola varies at a uniform rate r with respect to distances along the horizontal. Consequently, the grade diagram of a parabola is a straight line having a slope r.

2. For any curve, the difference between the ordinates at any two points on the curve is equal to the area under the grade diagram between these points.

Applying the second principle to Fig. A-1, we have

$$\Delta y = \text{area } cdmn = \left(\frac{G_c + G_d}{2}\right)\Delta x$$

and Eq. (17-9) is thus established. Now draw vertical lines through V and D in Fig. A-1*a*, and resolve the grade diagram into two parts by drawing the horizontal line *no*. Then

$$om = G_d - G_c = r\,\Delta x$$
$$D'D = \text{area } cdmn \qquad D'P = G_c\,\Delta x = \text{area } cdon$$
$$PD = D'D - D'P = \text{area } nom = \tfrac{1}{2}(r\,\Delta x)\,\Delta x$$
$$= \tfrac{1}{2}r(\Delta x)^2$$

The distance PD is the vertical offset of D from the tangent through C, and Eq. (17-10) is thus established.

In Fig. A-1*a*, VM is the vertical offset of M from the tangent through C and from the tangent through D. Applying Eq. (17-10), we have

$$VM = \tfrac{1}{2}r(CV')^2 = \tfrac{1}{2}r(V'D')^2$$
$$\therefore CV' = V'D'$$

and Eq. (17-8) is thus established.

Although Eq. (17-14) is a direct consequence of the tangent-offset method of curve plotting, it can readily be derived by means of the grade diagram, in this manner: In Fig. A-1,

$$\Delta y = \text{area } cdmn = \text{area } cdon + \text{area } nom$$
$$= G_c\,\Delta x + \tfrac{1}{2}r(\Delta x)^2$$

If C lies at the origin, Δx and Δy are replaced with x and y, respectively, and Eq. (17-14) is thus established.

We shall now derive Eq. (17-15). In Fig. A-2, AB is a parabolic arc having the same form as that in Fig. A-1*a*, and Q is an arbitrary point on this arc. Place the origin of coordinates at V, the point of intersection of the tangents through A and B, and let x and y denote the coordinates of Q. Draw a vertical line through Q, and let t_{QA} and t_{QB} denote the vertical offset of Q from the tangents through A and B, respectively. By Eq. (17-10),

$$t_{QA} = \tfrac{1}{2}r(L/2 + x)^2 = \frac{r(L + 2x)^2}{8}$$

$$t_{QB} = \tfrac{1}{2}r(L/2 - x)^2 = \frac{r(L - 2x)^2}{8}$$

But $\qquad t_{QA} = FQ = y - EF = y - G_a x$

and $\qquad t_{QB} = GQ = y - EG = y - G_b x$

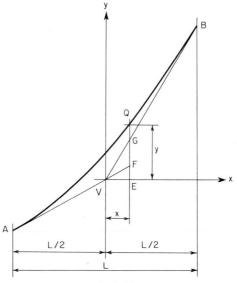

FIG. A-2

Then
$$y - G_a x = \frac{r(L + 2x)^2}{8} \qquad (a)$$

and
$$y - G_b x = \frac{r(L - 2x)^2}{8} \qquad (b)$$

Dividing Eq. (*a*) by Eq. (*b*), we obtain

$$\frac{y - G_a x}{y - G_b x} = \left(\frac{L + 2x}{L - 2x}\right)^2$$

and Eq. (17-15) follows at once.

APPENDIX B

ENGINEERING UNITS IN THE
INTERNATIONAL SYSTEM (SI)

The International System of Units is a modernized form of the metric system. It provides a set of units, a notational system for representing the units by symbols, and a set of prefixes that serve as multipliers of the units. The International System is abbreviated as SI from its French name, Le Système International d'Unités.

Classes of Units. In the International System, units are divided into the following categories:
1. Base units
2. Supplementary units
3. Derived units
   a. Derived units with special names
   b. Compound units

The seven *base* units, which are considered dimensionally independent, and the two *supplementary* units, which may be considered base units *or* derived units, serve as the building blocks of SI. These units and their respective symbols are recorded in Table B-1. *Derived* units are those that are formed by combining base units, supplementary units, or other derived units. Seventeen derived units

355

Table B-1. Base and Supplementary Units

Quantity	Unit	Symbol
Base units		
Length	meter	m
Mass	kilogram	kg
Time	second	s
Electric current	ampere	A
Thermodynamic temperature	kelvin	K
Amount of substance	mole	mol
Luminous intensity	candela	cd
Supplementary units		
Plane angle	radian	rad
Solid angle	steradian	sr

are assigned individual names and symbols; other derived units, called *compound* units, acquire their names and symbols from the mode of calculation. Derived units with special names and frequently used compound units are recorded in Tables B-2 and B-3, respectively. We shall define the first five derived units with special names.

Table B-2. Derived Units with Special Names*

Quantity	Unit	Symbol	Formula
Force	newton	N	$kg \cdot m/s^2$
Stress	pascal	Pa	N/m^2
Work, energy, quantity of heat	joule	J	$N \cdot m$
Power	watt	W	J/s
Frequency	hertz	Hz	$1/s$
Quantity of electricity	coulomb	C	$A \cdot s$
Electric potential	volt	V	W/A
Capacitance	farad	F	C/V
Electric resistance	ohm	Ω	V/A
Conductance	siemens	S	A/V
Magnetic flux	weber	Wb	$V \cdot s$
Magnetic flux density	tesla	T	Wb/m^2
Inductance	henry	H	Wb/A

* The other four derived units with special names—lumen (lm; ca·sr), lux (lx; lm/m²), becquerel (Bq; 1/s), and gray (gy; J/kg)—are not often encountered in structural engineering.

Table B-3. Compound Derived Units

Quantity	Unit	Symbol
Acceleration	meter per second squared	m/s^2
Angular acceleration	radian per second squared	rad/s^2
Angular velocity	radian per second	rad/s
Area	square meter	m^2
Entropy	joule per kelvin	J/K
Mass density	kilogram per cubic meter	kg/m^3
Moment of force	newton meter	$N \cdot m$
Specific energy	joule per kilogram	J/kg
Specific entropy	joule per kilogram kelvin	$J/(kg \cdot K)$
Specific volume	cubic meter per kilogram	m^3/kg
Velocity	meter per second	m/s
Viscosity, dynamic	pascal second	$Pa \cdot s$
Viscosity, kinematic	square meter per second	m^2/s
Volume	cubic meter	m^3

A *newton* is the force that must be applied to a body having a mass of 1 kilogram to give that body an acceleration of 1 meter per second squared.

A *pascal* is the stress or pressure caused by a force of 1 newton uniformly distributed over an area of 1 square meter.

A *joule* is the amount of work performed by a force of 1 newton when its point of application is displaced 1 meter in the direction of the force.

A *watt* is the power that generates 1 joule of work in 1 second.

A *hertz* is the frequency of a regularly recurring process in which 1 cycle is completed in 1 second.

As the foregoing definitions indicate, SI is a *coherent* system in the respect that the unit value of a physical quantity is defined exclusively in terms of the unit values of other physical quantities.

The compound derived units are assigned names and symbols that correspond to the manner in which their respective quantities are calculated. In forming the symbols, multiplication is indicated by means of a dot and division by use of a slant-rule fraction. For example, the symbol for newton meter is N·m and the symbol for joule per kilogram is J/kg. Powers of a unit are shown in exponent from rather than with abbreviations. For example, the symbol for volume is m^3 rather than cu m.

To illustrate the procedure for forming the names and symbols of the compound units, consider the quantity linear velocity, which is calculated in the following manner:

$$\text{Velocity} = \frac{\text{distance}}{\text{time}} = \frac{\text{meters}}{\text{seconds}}$$

The unit of linear velocity is therefore the meter per second, and its symbol is m/s. Similarly, consider the quantity specific volume, which is the volume occupied by a unit mass of a substance. Then

$$\text{Specific volume} = \frac{\text{volume}}{\text{mass}} = \frac{\text{cubic meters}}{\text{kilograms}}$$

The unit of specific volume is therefore the cubic meter per kilogram, and its symbol is m^3/kg.

Prefixes. By attaching prefixes to the units thus far considered, it becomes possible to form units that are decimal multiples or submultiples (decimal fractions) of the SI units. The motive in devising this system of multiples and submultiples is to permit all numerical quantities to be expressed in compact form. The SI prefixes and their respective symbols and numerical equivalents are recorded in Table B-4. As an illustration of their use, we have

$$1 \text{ kilonewton} = 1{,}000 \text{ newtons}$$

Table B-4. Prefixes*

Prefix	Symbol	Multiplication factor
exa	E	10^{18}
peta	P	10^{15}
tera	T	10^{12}
giga	G	10^{9}
mega	M	10^{6}
kilo	k	10^{3}
milli	m	10^{-3}
micro	μ	10^{-6}
nano	n	10^{-9}
pico	p	10^{-12}
femto	f	10^{-15}
atto	a	10^{-18}

* The prefixes hecto (h; 10^2), deka (da; 10^1), deci (d; 10^{-1}), and centi (c; 10^{-2}) should be avoided whenever possible.

or in symbolic form

$$1 \text{ kN} = 1{,}000 \text{ N}$$

Similarly,

$$1 \text{ mm} = 0.001 \text{ m} \qquad 1 \text{ MPa} = 1\ 000\ 000 \text{ Pa}$$

Multiples and submultiples must be formed in accordance with certain clearly defined rules, among which are the following:

1. For a unit of mass, the prefix is to be attached to the word *gram.*

2. Where the symbol of the compound unit is in the form of a fraction, the prefix is normally to be attached to the numerator. (An exception is the kilogram, which is a base unit.)

3. Where an exponent is present, it is understood that the entire unit *with its prefix* is to be raised to the indicated power. For example, the symbol mm^3 means cubic millimeter or millimeter cubed, and

$$1 \text{ mm}^3 = (10^{-3} \text{ m})^3 = 10^{-9} \text{ m}^3$$

It is recommended that prefixes be selected to restrict all numerical values to the range from 0.1 to 1000 wherever feasible.

Example B-1. Using multiples or submultiples, express the following values in the recommended manner: 21,400 N, 0.05 m/s^2, 118 × 10^6 Pa, and 720,000 m^2.

Solution

$$21{,}400 \text{ N} = 21.4 \times 10^3 \text{ N} = 21.4 \text{ kN}$$
$$0.05 \text{ m/s}^2 = 50 \times 10^{-3} \text{ m/s}^2 = 50 \text{ mm/s}^2$$
$$118 \times 10^6 \text{ Pa} = 118 \text{ MPa}$$
$$720{,}000 \text{ m}^2 = 0.72 \times 10^6 \text{ m}^2 = 0.72(10^3)^2 \text{ m}^2 = 0.72 \text{ km}^2$$

Conversion of Units.   In transferring numerical data from U.S. Customary units to the corresponding SI units, it is merely necessary to apply the appropriate conversion factors, and Table B-5 presents several such factors. To make these conversion factors more meaningful, we shall develop the relationship between 1 lb of force and 1 N. Since the metric system regards mass as a basic property and force as a derived property, it is imperative that we distinguish clearly between the mass and weight of a body.

Table B-5. Conversion Factors

To convert from	To	Multiply by
Acre	meters^2	$4.046\ 873 \times 10^3$
British thermal unit		
(International Table)	joule	$1.055\ 056 \times 10^3$
Cubic foot (or foot^3)	meter^3	$2.831\ 685 \times 10^{-2}$
Cubic inch (or inch^3)	meter^3	$1.638\ 706 \times 10^{-5}$
Degree (of angle)	radian	$1.745\ 329 \times 10^{-2}$
Degree Rankine	kelvin	$1/1.8$
Foot	meter	$3.048\ 000 \times 10^{-1}$
Foot-pound	joule	$1.355\ 818$
Gallon (U.S. liquid)	meter^3	$3.785\ 412 \times 10^{-3}$
Horsepower		
(550 ft-lb/sec)	watt	$7.456\ 999 \times 10^2$
Inch	meter	$2.540\ 000 \times 10^{-2}$
Pound of force	newton	$4.448\ 222$
Pound of mass	kilogram	$4.535\ 924 \times 10^{-1}$
Pound per square foot	pascal	$4.788\ 026 \times 10$
Pound per square inch	pascal	$6.894\ 757 \times 10^3$
Square foot	meter^2	$9.290\ 304 \times 10^{-2}$
Square inch	meter^2	$6.451\ 600 \times 10^{-4}$

Note: The term *pound* refers to force except as noted.

According to Newton's second law of motion, a single force acting on a body imparts to the body an acceleration that is directly proportional to the force. However, the constant of proportionality has a unique value for each body, and this constant of proportionality is termed the *mass* of the body. Thus, mass is an intrinsic property of a body, being the ratio of the force acting on the body to the resulting acceleration. For example, if a body is subjected to a single force of 240 N and thereby acquires an acceleration of 5 m/s^2, the mass of the body is $240/5 = 48$ kg. The "weight" of a body is the force of attraction that the earth (or other planet) exerts on the body. Thus, weight is an extrinsic property, for its value varies according to the location of the body. For example, a body that weighs precisely 1000 lb at the Panama Canal weighs 1004.4 lb in Greenland.

The metric system has three units at its core: the meter, kilogram, and second, these being the units of length, mass, and time, respectively. The kilogram was intended to be the mass of 1000 cu cm of pure water at 4°C, which weighs approximately 2.205 lb at the standard station. If a 1-lb mass is defined as the mass of a body that weighs 1 lb at the standard station, it follows that 2.205 lb of mass corresponds to 1 kg, or 1 lb of mass corresponds to 1/2.205 or 0.4535 kg. Now, 1 lb of force will give a body of 1-lb mass an acceleration of 9.807 m/s². Equating force to the product of mass and acceleration, we obtain the following approximations:

$$1 \text{ lb of force} = (1 \text{ lb of mass})(9.807 \text{ m/s}^2)$$
$$= (0.4535 \text{ kg})(9.807 \text{ m/s}^2)$$
$$= 4.447 \text{ kg} \cdot \text{m/s}^2$$

But
$$1 \text{ N} = 1 \text{ kg} \cdot \text{m/s}^2$$

Thus, 1 lb of force corresponds approximately to 4.447 N.

Conversion factors corresponding to more complex units may readily be obtained by combining the conversion factors shown in Table B-5. For example, assume that the volumetric flow rate of a liquid is expressed in the unit cfm (cubic feet per minute), and that a transfer is to be made to the corresponding SI unit. Then

$$\text{cfm} = \frac{\text{ft}^3}{\text{min}} = \frac{\text{ft}^3}{60 \text{ s}} = \frac{2.831\ 685 \times 10^{-2} \text{ m}^3}{60 \text{ s}}$$
$$= 0.04719 \times 10^{-2} \text{ m}^3/\text{s} = 4.719 \times 10^{-4} \text{ m}^3/\text{s}$$

Example B-2. Water discharges over a dam at the rate of 1600 cfm. What is the discharge in m³/s, to three significant figures?

Solution

$$Q = 1600 \times 4.719 \times 10^{-4} = 7550 \times 10^{-4} \text{ m}^3/\text{s} = \textbf{0.755 m}^3\textbf{/s}$$

If the numerical data in a given problem are presented in U.S. Customary units and the answer is to be expressed in an SI unit, only the answer itself requires conversion, as the following example illustrates.

Example B-3. An open concrete-lined canal of rectangular cross section is 12 ft wide and 9 ft deep, and its elevation drops at the rate of 1.60 ft per mile. If the canal is flowing full, what is the rate at which it is discharging water, in m³/s?

Solution.   Refer to Art. 16-7. We shall apply Eq. (16-9) with $n = 0.016$.

$$A = 12 \times 9 = 108 \text{ sq ft} \qquad WP = 12 + 2 \times 9 = 30 \text{ ft}$$

$$R = \frac{108}{30} = 3.6 \text{ ft} \qquad\qquad s = \frac{1.60}{5280} = 0.000303$$

$$V = (1.486/0.016) \times 3.6^{2/3} \times 0.000303^{1/2}$$
$$= (1.486/0.016) \times 2.35 \times 0.0174 = 3.80 \text{ fps}$$
$$Q = AV = 108 \times 3.80 = 410 \text{ cfs}$$

Applying the conversion factor given in Table B-5, we have

$$Q = 410 \times 2.832 \times 10^{-2} = \textbf{11.6 m}^3\textbf{/s}$$

For our present purpose, it is convenient to divide all engineering equations into two broad categories, in this manner:

1. Equations that are independent of the system of units in which the quantities are expressed.

2. Equations that are dependent on the system of units because they contain a number that is governed by the system.

The first category is illustrated by Eq. (5-5) for the maximum bending moment in a simply supported beam carrying a uniformly distributed load. An equation of this type requires merely that one system of units be applied consistently.

Example B-4.   A beam on a simple span of 6 m has an allowable bending moment of 150 kN·m. If the beam is to carry a uniformly distributed load, what is the allowable unit load?

Solution.   Solving Eq. (5-5) for w, we have

$$w = \frac{8M}{L^2} = \frac{8 \times 150}{6^2} = \textbf{33.3 kN/m}$$

The second category of equations is illustrated by Eq. (16-6b),

$$Q = 3.33(L - 0.2h)h^{3/2} \qquad \text{cfs}$$

where Q is the volumetric discharge over a sharp-edged rectangular weir having two end contractions. We shall refer to the first term at the right as the *numerical quantity*. The value of this quantity is governed by the system of units in which the equation is expressed, and the value of 3.33 is valid solely with respect to the ft-sec system. To determine the value of the numerical quantity under SI, it is necessary to identify the unit that is associated with this quantity. Let U denote the unit under the ft-sec system. Writing Eq. (16-6b) in dimensional form (i.e., in terms of units rather than numerical quantities), we have

$$\frac{ft^3}{sec} = U \times ft \times ft^{3/2}$$

Solving,
$$U = \frac{ft^{1/2}}{sec}$$

When this unit is converted to SI, the value of the numerical quantity becomes

$$3.33 \times \frac{0.3048^{1/2}}{1} = 1.84$$

and Eq. (16-6b) is transformed to

$$Q = 1.84(L - 0.2h)h^{3/2} \qquad m^3/s \qquad \text{(B-1)}$$

Example B-5. Applying the data presented in Example 16-6, compute the discharge in m^3/s (a) by converting the calculated value of Q only; (b) by converting the values of L and h and then applying Eq. (B-1).

Solution. The given values are

$$L = 10 \text{ ft} \qquad h = 9 \text{ in.} = 0.75 \text{ ft}$$

PART a: From Example 16-6, $Q = 21.3$ cfs. Then

$$Q = 21.3 \times 2.832 \times 10^{-2} = \mathbf{0.603 \ m^3/s}$$

PART b

$$L = 10 \times 0.3048 = 3.048 \text{ m}$$
$$h = 0.75 \times 0.3048 = 0.2286 \text{ m}$$

$$Q = 1.84(3.048 - 0.2 \times 0.2286)0.2286^{3/2} = 0.604 \ m^3/s$$

BIBLIOGRAPHY

In addition to the following list of books, the reader should consult the catalogs of the various organizations that publish books and technical papers in civil engineering. Among these organizations are the following: American Association of State Highway Officials (Washington, D.C.), American Concrete Institute (Detroit), American Institute of Steel Construction (New York), American Public Health Association (New York), American Railway Engineering Association (Chicago), American Society of Civil Engineers (New York), Concrete Reinforcing Steel Institute (Chicago), and Wire Reinforcement Institute (Washington, D.C.).

The engineer who wishes to keep abreast of developments in this dynamic and ever-expanding field will find the host of technical periodicals an invaluable source of information. The important engineering literature of the year is listed and briefly described in *The Engineering Index*, which is available for reference in public libraries.

GENERAL

Abbett, R. W.: *American Civil Engineering Practice*, Wiley, vols. I and II, 1956; vol. III, 1957.

Merritt, F. S.: *Standard Handbook for Civil Engineers*, McGraw-Hill, 1968.
Urquhart, L. C.: *Civil Engineering Handbook*, 4th ed., McGraw-Hill, 1959.

STATICS AND MECHANICS OF MATERIALS

Beer, F. P., and E. R. Johnston, Jr.: *Mechanics for Engineers*, vol. 1, 2d ed., McGraw-Hill, 1962.
Biggs, J. M.: *Introduction to Structural Dynamics*, McGraw-Hill, 1964.
Brown, G. W.: *Applied Mechanics*, Prentice-Hall, 1971.
Chajes, A.: *Principles of Structural Stability Theory*, Prentice-Hall, 1974.
Crandall, S. H., N. C. Dahl, and T. J. Lardner: *Introduction to the Mechanics of Solids*, 2d ed., McGraw-Hill, 1972.
Den Hartog, J. P.: *Advanced Strength of Materials*, McGraw-Hill, 1952.
Fitzgerald, R. W.: *Strength of Materials*, Addison-Wesley, 1967.
Griffel, W.: *Handbook of Formulas for Stress and Strain*, Frederick Ungar, 1966.
Higdon, A., E. H. Ohlsen, W. B. Stiles, and J. A. Weese: *Mechanics of Materials*, 2d ed., Wiley, 1967.
Jensen, A., and H. Chenoweth: *Applied Strength of Materials*, 3d ed., McGraw-Hill, 1975.
——— and ———: *Statics and Strength of Materials*, 2d ed., McGraw-Hill, 1967.
Meriam, J. L.: *Statics*, 2d ed., Wiley, 1971.
Roark, R. J.: *Formulas for Stress and Strain*, 4th ed., McGraw-Hill, 1965.
Seely, F. B., and J. O. Smith: *Advanced Mechanics of Materials*, 2d ed., Wiley, 1952.
Shames, I. H.: *Introduction to Statics*, Prentice-Hall, 1971.
Shanley, F. R.: *Mechanics of Materials*, McGraw-Hill, 1967.
Snyder, R. D., and E. F. Byars: *Engineering Mechanics: Statics and Strength of Materials*, McGraw-Hill, 1973.
Timoshenko, S. P., and J. Gere: *Theory of Elastic Stability*, 2d ed., McGraw-Hill, 1961.
——— and J. N. Goodier: *Theory of Elasticity*, 3d ed., McGraw-Hill, 1970.
——— and D. H. Young: *Elements of Strength of Materials*, 5th ed., Van Nostrand Reinhold, 1968.

STRUCTURAL DESIGN

American Institute of Timber Construction: *Timber Construction Manual*, 2d ed., Wiley, 1974.
Amrhein, J. E.: *Reinforced Masonry Engineering Handbook*, 2d ed., Masonry Institute of America, Los Angeles, 1972.
Beedle, L. S.: *Plastic Design of Steel Frames*, Wiley, 1958.
Borg, S. F., and J. J. Gennaro: *Modern Structural Analysis*, Van Nostrand Reinhold, 1969.
Desch, H. E.: *Timber: Its Structure and Properties*, 5th ed., St. Martin's Press, 1973.
Ferguson, P. M.: *Reinforced Concrete Fundamentals*, 3d ed., Wiley, 1973.
Fintel, M.: *Handbook of Concrete Engineering*, Van Nostrand Reinhold, 1974.
Gaylord, E. H., and C. N. Gaylord: *Design of Steel Structures*, 2d ed., McGraw-Hill, 1972.

———— and ————: *Structural Engineering Handbook*, McGraw-Hill, 1968.

Gere, J. M., and W. Weaver, Jr.: *Analysis of Framed Structures*, Van Nostrand Reinhold, 1965.

Hoadley A.: *Essentials of Structural Design*, Wiley, 1964.

Hodge, P. G.: *Plastic Analysis of Structures*, McGraw-Hill, 1959.

Horne, M. R.: *Plastic Theory of Structures*, M.I.T. Press, 1972.

Kurtz, M.: *Comprehensive Structural Design Guide*, McGraw-Hill, 1969.

La Londe, W. S., Jr., and M. F. Janes: *Concrete Engineering Handbook*, McGraw-Hill, 1961.

Laursen, H. I.: *Structural Analysis*, McGraw-Hill, 1969.

Lincoln Electric Co.: *Procedure Handbook of Arc Welding Design and Practice*, 12th ed., Cleveland, 1973.

Merritt, F. S.: *Structural Steel Designers' Handbook*, McGraw-Hill, 1972.

Norris, C. H., et al.: *Structural Design for Dynamic Loads*, McGraw-Hill, 1959.

Rogers, P.: *Reinforced Concrete Design for Buildings*, Van Nostrand Reinhold, 1973.

Salmon, C. G., and J. E. Johnson: *Steel Structures: Design and Behavior*, Crowell, 1972.

Shermer, C. L.: *Design in Structural Steel*, Wiley, 1972.

Silvester, F. D.: *Mechanical Properties of Timber*, Pergamon Press, 1967.

Tall, L., et al.: *Structural Steel Design*. 2d ed., Ronald, 1974.

Timber Engineering Company: *Timber Design and Construction Handbook*, McGraw-Hill, 1956.

Timoshenko, S. P., and D. H. Young: *Theory of Structures*, 2d ed., McGraw-Hill, 1965.

Troxell, G. E., H. E. Davis, and J. W. Kelly: *Composition and Properties of Concrete*, 2d ed., McGraw-Hill, 1968.

U.S. Department of Agriculture, Forest Products Laboratory: *Wood Handbook* (Agriculture Handbook 72), Government Printing Office, 1974.

U.S. Department of the Interior, Bureau of Reclamation: *Design of Small Dams*, 2d ed., 1973.

Wang, C. K., and C. L. Eckel: *Elementary Theory of Structures*, McGraw-Hill, 1957.

Winter, G., L. C. Urquhart, C. E. O'Rourke, and A. H. Nilson: *Design of Concrete Structures*, 8th ed., McGraw-Hill, 1972.

FLUID MECHANICS

Daugherty, R. L., and J. B. Franzini: *Fluid Mechanics with Engineering Applications*, 6th ed., McGraw-Hill, 1965.

Davis, C. V., and K. E. Sorensen: *Handbook of Applied Hydraulics*, 3d ed., McGraw-Hill, 1969.

Fox, J. A.: *Engineering Fluid Mechanics*, McGraw-Hill, 1974.

Hansen, A. G.: *Fluid Mechanics*, Wiley, 1967.

Henderson, F. M.: *Open Channel Flow*, Macmillan, 1966.

Hicks, T. G., and T. Edwards: *Pump Application Engineering*, McGraw-Hill, 1971.

King, H. W., and E. F. Brater: *Handbook of Hydraulics*, 5th ed., McGraw-Hill, 1963.

Michelson, I.: *Science of Fluids*, Van Nostrand Reinhold, 1970.
Morris, H. M., and J. M. Wiggert: *Applied Hydraulics in Engineering*, 2d ed., Wiley, 1972.
Stepanoff, A. J.: *Centrifugal and Axial Flow Pumps*, 2d ed., Wiley, 1957.
Streeter, V. L., and B. Wylie: *Fluid Mechanics*, 6th ed., McGraw-Hill, 1975.

SURVEYING AND ROUTE DESIGN

Allen, C. F.: *Railroad Curves and Earthwork*, 7th ed., McGraw-Hill, 1931.
Brinker, R. C.: *Elementary Surveying*, 5th ed., Crowell, 1969.
Davies, R. E., F. S. Foote, and J. W. Kelly: *Surveying: Theory and Practice*, 5th ed., McGraw-Hill, 1966.
_____ and J. W. Kelly: *Elementary Plane Surveying*, 4th ed., McGraw-Hill, 1967.
Hallert, B.: *Photogrammetry: Basic Principles and General Survey*, McGraw-Hill, 1960.
Hickerson, T. F.: *Route Location and Design*, 5th ed., McGraw-Hill, 1967.
Kissam, P.: *Surveying for Civil Engineers*, McGraw-Hill, 1956.
_____: *Surveying Practice*, 2d ed., McGraw-Hill, 1971.
Meyer, C. F.: *Route Surveying and Design*, 4th ed., Crowell, 1969.
Nassau, J. J.: *Practical Astronomy*, 2d ed., McGraw-Hill, 1968.

SOIL MECHANICS AND FOUNDATIONS

Barkan, D. D.: *Dynamics of Bases and Foundations*, McGraw-Hill, 1962.
Chellis, R. D.: *Pile Foundations*, 2d ed., McGraw-Hill, 1961.
Dunham, C. W.: *Foundations of Structures*, 2d ed., McGraw-Hill, 1962.
Hough, B. K.: *Basic Soils Engineering*, 2d ed., Wiley, 1969.
Ritter, L. J., Jr., and R. J. Paquette: *Highway Engineering*, 3d ed., Ronald, 1967.
Spangler, M. G., and R. L. Handy: *Soil Engineering*, 3d ed., Crowell, 1973.
Terzaghi, K.: *Theoretical Soil Mechanics*, Wiley, 1943.
_____ and R. B. Peck: *Soil Mechanics in Engineering Practice*, 2d ed., Wiley, 1967.
Tschebotarioff, G. P.: *Foundations, Retaining and Earth Structures*, 2d ed., McGraw-Hill, 1973.

SANITARY ENGINEERING

Babbitt, H. E., J. J. Doland, and J. L. Cleasby: *Water Supply Engineering*, 6th ed., McGraw-Hill, 1962.
Ciaccio, L. L.: *Water and Water Pollution Handbook*, (4 vols.), Marcel Dekker, New York, 1971.
Fair, G. M., J. C. Geyer, and D. A. Okun: *Elements of Water Supply and Waste Water Disposal*, 2d ed., Wiley, 1971.
Imhoff, K., and G, Fair: *Sewage Treatment*, 2d ed., Wiley, 1956.

Metcalf, L., and H. P. Eddy: *Wastewater Engineering: Collection, Treatment, Disposal*, McGraw-Hill, 1972.

Sawyer, C. N., and P. L. McCarty: *Chemistry for Sanitary Engineers*, 2d ed., McGraw-Hill, 1967.

Steel, E. W.: *Water Supply and Sewerage*, 4th ed., McGraw-Hill, 1960.

Tebbutt, T. H. Y.: *Principles of Water Quality Control*, Pergamon Press, 1971.

INDEX

Aerobic stabilization, 345
Affinity laws, 287
Angle:
  of contact, 28
  deflection, 290
  of friction, 25
  intersection, 290
  of repose, 26
  of twist, 164–165
Arc of contact, 28
Area:
  calculation of, 313–316
  centroid of, 36
Astronomy, field, 319–321
Average-end-area method, 309–311
Average-grade method, 297–299
Axis:
  centroidal, 35–37
  longitudinal, 48
  prinicipal, 178
Azimuth of star, 320

Beam:
  with axial loading, 123–126
  bending moment in, 95–103

Beam (*Cont.*):
  bending stress in, 108–114
  composite, 120–123
  continuous, 155–160
  definition of, 92
  deflection of, 148–154, 160–161
  horizontal shear in, 114–120
  moving loads on, 103–108
  neutral axis of, 109–110
  reactions for, 93–95
  reinforced-concrete (*see*
    Reinforced-concrete beam)
  rigidity of, 154–155
  shearing stresses in, 114–120
  statically indeterminate, 155–160
  steel (*see* Steel beam)
  with three reactions, 155–157
  timber, 208–209
  with torsion, 171–173
  types of, 93
  vertical shear in, 95–103
Beam-column, steel, 182–184
Beam-shaft, 171–173
Bending moment:
  in beam, 95–103
  in continuous beam, 157–160, 252

369

Bending moment (*Cont.*):
    definition of: in beam, 95–97
        in bridge truss, 195
Bending-moment diagram:
    construction of, 97–103
    properties of, 98–100
    for single moving load, 103–104
Bending stress:
    allowable, in steel beam, 133–136
    in beam, 108–114
    calculation of, 108–114
    in column footing, 269
Bernoulli's theorem, 274–276
Biochemical oxygen demand (BOD), 346
BOD (biochemical oxygen demand), 346
Borrow material, 328–330
Bow's notation, 192
Bridge truss, 194–205
Bulking of sand, 231
Buoyancy, 273–274

Centrifugal pump, 286–288
Centroid of area, 36
Centroidal axis, 35–37
Channel, flow in, 283–286
Chézy formula, 283
Circle of influence, 341
Coefficient:
    of deoxygenation, 347
    of permeability, 342
    roughness, 284
    of static friction, 25
    of thermal expansion, 67
Cohesion of soil, 333–334
Column:
    effective length of, 177
    reinforced-concrete (*see*
        Reinforced-concrete column)
    steel: under axial load, 177–182
        with moment, 182–184
    timber, 211–213
Column footing, 267–272
Compact steel section, 133
Composite beam, 120–123
Composite member, 49
Compound shaft, 166–168
Compression index of soil, 338

Compression member:
    under bending, 123–126
    timber, under oblique force, 209–211
Compression test for soil, 334–337
Concrete:
    properties of, 226–231
    slump test of, 229
    ultimate strength of, 226
    (*See also* Reinforced-concrete beam;
        Reinforced-concrete column)
Concrete mixture, 227–231
Cone of depression, 341
Contact:
    angle of, 28
    arc of, 28
Continuous beam:
    bending moment in (*see* Bending
        moment, in continuous beam)
    reactions for, 155–160
    of reinforced concrete, 252–256
Couple, force, 4–5
Cover plates:
    for plate girder, 138
    for steel beam, 136–138
Cross section, definition of, 48
Culmination of Polaris, 319–320
Curve:
    degree of, 289
    elastic (*see* Elastic curve)
    horizontal circular: plotting of,
        289–294
        sight distance of, 307
    parabolic: grade diagram of, 301
        properties of, 295–297, 351–354
    vertical (*see* Vertical parabolic
        curve)

Dam, gravity, 219–221
Darcy-Weisbach formula, 278–280
Deficit, oxygen, 347
Deflection of beam, 148–154, 160–161
Deflection angle, 290
Deformation:
    of axially loaded body, 59–62
    elastic, 59
    plastic, 59
    of pressure vessel, 78
    of shaft, 164–168

Degree:
  of curve, 289
  of saturation, 323
Deoxygenation, coefficient of, 347
Departure of line, 311–313
Development length, 244–247
Differential leveling, 316–317
Distance:
  double meridian, 313–314
  sight, 307–309
Distributed force, resultant of, 39–40
Double-integration method, 151–154
Double meridian distance, 313–314
Dowel in column footing, 272
Drawdown of well, 341

Earth pressure (*see* Soil pressure)
Earthwork, volume of, 309–311
Eccentric load:
  on pile group, 216–218
  on rectangular member, 123–126
  on reinforced-concrete column,
    264–266
  on riveted joint, 168–171
Effective length of column, 177
Elastic curve:
  curvature of, 149–151
  definition of, 148
Elastic deformation, 59
Elastic design, 71–72
Elastic limit, 59
Elasticity, modulus of, 60
  shearing, 62
Equilibrium, conditions of, 14

Factor, load, 71
Field astronomy, 319–321
First moment (*see* Statical moment)
Flexural stress (*see* Bending stress)
Flexure formula, 110
Fluid, power of, 276–277
Fluid flow, 274–288
Footing, column, 267–272
Force(s):
  components of, 6
  composition of, 11
  definitions pertaining to, 1–2
  internal, 17

Force(s) (*Cont*.):
  moment of, 4–5
  resolution of, 6
  system of (*see* Force system)
  torque of, 4–5
  types of, 15, 49
Force polygon, 13
Force system:
  balanced, 14
  characteristics of, 2–5
  resultant of, 11–13
  transformation of, 5–11
Forest Products Laboratory formulas,
    211–212
Francis formula, 278
Free-body diagram, 17
Friction:
  angle of, 25
  belt, 28–30
  in fluid flow, 278–280
  in soil, 333–334
  static, 24–28
    coefficient of, 25

Grade diagram, 301, 351–354
Graphical analysis of truss, 191–193
Gravity dam, 219–221
Gravity structure, 219
Gyration, radius of, 41–42

Hankinson's formula, 209
Hoop, circular, 68–70
Hoop stress, 70, 76–78
Hoop tension, 70, 76–78
Horizontal circular curve (*see* Curve,
    horizontal circular)
Horizontal shear in beam, 114–120
Horsepower, 276
Hydraulic radius, 283

Indeterminate axial member, 62–66
Indeterminate beam, 155–160
Indeterminate shaft, 166–168
Inertia:
  moment of, 40–44
  polar, 44–45
  product of, 44–45

Influence line, 196–201
Inlet time, 344
Interaction diagram, 260–264
International System of Units (SI),
    355–363
Intersection angle, 290

Joint(s):
    method of, 187
    riveted (*see* Riveted joint)
    types of, 16
    welded, 88–90

Kern of rectangular section, 126
Kip, definition of, 2
Kutter formula, 284

Latitude of line, 311–313
Leveling, differential, 316–317
Limit:
    elastic, 59
    proportional, 60
    of soil states, 326–328
Line of action, 1
Load(s):
    definition of, 15
    eccentric (*see* Eccentric load)
    moving: on beam, 103–108
        on bridge truss, 196–205
    types of, 15
    ultimate, 233
Load factor, 71
Longitudinal axis, 48

Manning formula, 284
Method of sections and joints, 187
Middle third, principle of, 125
Mixture:
    concrete, 227–231
    soil, 330–333
Modular ratio, 120
Modulus:
    of elasticity, 60
        shearing, 62
    of rigidity, 62
    section, 110

Modulus (*Cont.*):
    shearing, of elasticity, 62
    Young's, 60
Mohr's circle of stress:
    construction of, 53–58
    for soil mass, 333–337
Moisture content of soil, 323
Moment:
    on base of structure, 214
    bending (*see* Bending moment)
    of force, 4–5
    of inertia, 40–44
        polar, 44–45
    statical (*see* Statical moment)
    ultimate, 233
Moment diagram (*see* Bending-
    moment diagram)
Moving loads (*see* Loads, moving)

Neutral axis of beam, 109–110
Neutral point, 197

Orifice, flow through, 277
Oxygen deficit, 347
Oxygen demand, biochemical (BOD),
    346

Parabolic curve (*see* Curve, parabolic;
    Vertical parabolic curve)
Permeability, coefficient of, 342
Pile group, eccentrically loaded,
    216–218
Pipe:
    flow through, 278–283
    prestressed concrete, 78–82
Planes, principal, 55–58
Plastic deformation, 59
Plastic design, 71
Plasticity index, 326–328
Plate girder, 138–147
    cover plates for, 138
    procedure for designing, 139
    rivet pitch in, 145–147
    stiffeners for, 141–145
Point, neutral, 197
Poisson's ratio, 61

Polar moment of inertia, 44–45
Polaris, culmination of, 319–320
Porosity of soil, 323
Power of fluid, 276–277
Pressure, 39, 49
    soil (*see* Soil pressure)
Prestressing of pipe, 78–82
Principal axes, 178
Principal planes and stresses, 55–58
Principle:
    of middle third, 125
    of transmissibility, 2–3
Prismatic member, 48
Prismoidal method, 309–311
Product of inertia, 44–45
Proportional limit, 60
Pump, centrifugal, 286–288

Radius of gyration, 41–42
Rainfall, intensity of, 344
Rankine formulas, 222–224
Rational method, 344–345
Reaction, definition of, 15
Reaeration of water, 347
Reinforced-concrete beam, 231–256
    balanced, 233
    basic characteristics of, 231
    bond stress in, 244–247
    continuous, 252–256
    doubly reinforced, 241–244
    failure of, 233–234
    notation for, 232
    proportions of, 231
    rectangular, 234–238, 241–244
    shearing stress in, 244–245
    stirrup design for, 247–251
    of T shape, 238–241
Reinforced-concrete column, 257–266
    eccentrically loaded, 264–266
    equations for, 259–260
    failure of, 257
    interaction diagram for, 260–264
    notation for, 258
    types of, 257
Reinforcing bars:
    development of, 244–247
    sizes of, 226–227
Relative density of soil, 326–328
Repose, angle of, 26

Resultant:
    of distributed force, 39–40
    of force system, 11–13
Retaining wall, 221–224
Rigidity:
    of beam, 154–155
    modulus of, 62
Rivet pitch in plate girder, 145–147
Riveted joint, 82–88, 168–171
    eccentrically loaded, 168–171
    efficiency of, 85
    stresses in, 83–88
    torsion on, 168–171
    types of, 82
Roof truss, loading of, 193–194
Roughness coefficient, 284

Saturation, degree of, 323
Saturation content of water, 347
Second moment (*see* Moment of
    inertia)
Section(s):
    cross, 48
    method of, 187
    transformed, 121–123
Section modulus, definition of, 110
Self-purification constant, 348
Settlement of structures, 337–340
Sewage, stabilization of, 345–350
Shaft, 162–168
    beam-shaft, 171–173
    compound, 166–168
    deformation of, 164–168
    shearing stress in, 162–168
    statically indeterminate, 166–168
    torsion of, 162–168
Shear:
    in column footing, 268
    horizontal, 114–120
    pure, 57–58
    vertical (*see* Vertical shear)
Shear diagram (*see* Vertical-shear
    diagram)
Shear flow, 115
Shearing modulus of elasticity, 62
Shearing stress(es):
    in beam, 114–120
    on collinear planes, 51–58
    in column footing, 268–272

Shearing stress(es) (*Cont.*):
  definition of, 49
  in reinforced-concrete beam,
    244–245
  in shaft, 162–168
  in steel beam, 131–133
SI (International System of Units),
  355–363
Sieve analysis, 330–333
Sight distance, 307–309
Simpson's rule, 315–316
Slenderness ratio, 178
Slump test of concrete, 229
Soil:
  cohesion of, 333–334
  composition of, 322–328
  compressibility of, 337–340
  compression index of, 338
  compression test for, 334–337
  friction in, 333–334
  moisture content of, 323
  porosity of, 323
  relative density of, 326–328
  stresses in, 333–337
  void ratio of, 323
Soil mixture, 330–333
Soil pressure:
  active and passive, 224
  calculation of, 214–216, 221–224
Specific capacity of well, 342
Specific speed, 287–288
Spiral, use of, 294
Stability:
  criteria for, 218–219
  of triangle, 185–186
Stabilization, aerobic, 345
Stadia surveying, 318–319
Star, azimuth of, 320
Static friction, 24–28
Statical moment:
  of area, 34–39
  of partial area, 45–46
Statically determinate structure, 19
Statically indeterminate axial member,
  62–66
Statically indeterminate beam, 155–160
Statically indeterminate shaft, 166–168
Steel beam, 131–138
  allowable bending stress in, 133–136
  cover plates for, 136–138

Steel beam (*Cont.*):
  moment resistance of, 133
  shearing stress in, 131–133
  vertical shear in, 131–132
Steel beam-column, 182–184
Steel column (*see* Column, steel)
Stiffeners for plate girder, 141–145
Stirrups, 247–251
Storm drain, design of, 344–345
Strain:
  axial, 59–61
  shearing, 62
Strength design, 71–72
Stress(es):
  bending (*see* Bending stress)
  bond, 244–247
  on collinear planes, 51–58
  definition of, 49
  hoop, 70, 76–78
  Mohr's circle of, 53–58
  in pressure vessel, 76–78
  principal, 55–58
  shearing [*see* Shearing stress(es)]
  sign convention for, 52
  in soil mass, 333–337
  state of, 51
  thermal, 66–69
  types of, 49
  yield-point, 61, 71, 131
Support, types of, 16

T beam, reinforced-concrete, 238–
  241
Talbot's formula, 344
Tangent offset, 296–297
Tangent-offset method, 297–300
Temperature reinforcement, 256
Tension-field action, 143
Tension member, design of, 174–177
Terzaghi-Peck formula, 338
Theorem of three moments, 157–160
Thermal expansion, coefficient of, 67
Thermal stress, 66–69
Three-moments, theorem of, 157–160
Timber, 208–213
  characteristics of, 207
Timber beam, 208–209
Timber column, 211–213

Timber compression member, obliquely loaded, 209–211
Time of concentration, 344
Torque of force, 4–5
Torsion:
  beam with, 171–173
  on rivet group, 168–171
  of shaft, 162–168
Transformed section, 121–123
Transmissibility, principle of, 2–3
Trapezoidal rule, 315–316
Traverse, closed, 311–313
Triangle, stability of, 185–186
Truss, 185–206
  analysis of, 187–193
    graphical, 191–193
  bridge, 194–205
  definition of, 186–187
  roof, loading of, 193–194
Twist, angle of, 164–165

Ultimate load and moment, 233
Ultimate strength of concrete, 226
Ultimate-strength design, 71–72
Units, International System of (SI), 355–363

Vector representation, 2
Vertical parabolic curve, 294–309
  to contain given point, 305–306

Vertical parabolic curve (*Cont.*):
  grade diagram of, 301
  plotting of, 297–307
  sight distance of, 308–309
Vertical shear:
  in beam, 95–103
  definition of: in beam, 95–97
    in bridge truss, 195
    in steel beam, 131–132
Vertical-shear diagram:
  construction of, 97–103
  properties of, 98–100
Void ratio of soil, 323

Water:
  reaeration of, 347
  saturation content of, 347
Water table, 325
Weir, flow over, 277–278
Welded joint, 88–90
Well:
  drawdown of, 341
  hydraulics of, 341–343
  specific capacity of, 342
Wetted perimeter, 283
Wood (*see* Timber)
Working-stress design, 71

Yield-point stress, 61, 71, 131
Young's modulus, 60